Innovating Science Teacher Education

"This is an important study. Science teaching and the preparation of science teachers is dominated by a far too uncomplicated understanding of the nature of science. Mansoor Niaz brings a strong and clear mastery of the history and philosophy of science to bear on pressing issues in the teaching of science. He presents a valuable perspective on how we should understand the nature of science and how we can work with pre-service and in-service teachers to strengthen their appreciation."

Louis Rosenblatt, Baltimore Freedom Academy

Science does not advance by just doing experiments and collecting data. Progress in science inevitably leads to controversies and alternative interpretations of data. How teachers view the nature of scientific knowledge is crucial to their understanding of science content and how it can be taught.

This book presents an overview of the dynamics of scientific progress and its relationship to the history and philosophy of science, and then explores their methodological and educational implications and develops innovative strategies based on actual classroom practice for teaching topics such as the nature of science, conceptual change, constructivism, qualitative-quantitative research, and the role of controversies, presuppositions, speculations, hypotheses, and predictions.

In recent decades a worldwide sustained effort has been underway to introduce history and philosophy of science into the science curriculum, textbooks, and classrooms. Implementation of these reform projects requires teacher training that promotes an understanding of the nature of science and the dynamics of scientific progress. Field-tested in science education courses, the book is designed to involve readers in critically thinking about history and philosophy of science and to engage science educators in learning how to progressively introduce various aspects of "science-in-the-making" in their classrooms, to promote discussions highlighting controversial historical episodes included in the science curriculum, and to expose their students to the controversies and encourage them to support, defend, or critique the different interpretations. *Innovating Science Teacher Education* offers guidelines to go beyond traditional textbooks, curricula, and teaching methods and innovate with respect to science teacher education and classroom teaching.

Mansoor Niaz is Professor at the Chemistry Department, Universidad de Oriente, Cumaná, Venezuela.

Innovating Science Teacher Education

A History and Philosophy of Science Perspective

Mansoor Niaz

NEW YORK AND LONDON

KH

First published 2011
by Routledge
270 Madison Avenue, New York, NY 10016

Simultaneously published in the UK
by Routledge
2 Park Square, Milton Park, Abingdon, Oxon OX14 4RN

Routledge is an imprint of the Taylor & Francis Group, an informa business

Typeset in Minion by Wearset Ltd, Boldon, Tyne and Wear
Printed and bound in the United States of America on acid-free paper by Walsworth Publishing Company, Marcelino, MO.

Library of Congress Cataloging-in-Publication Data
Niaz, Mansoor.
Innovating science teacher education: a history and philosophy of science perspective/Mansoor Niaz.
p. cm.
Includes bibliographical references and index.
1. Science–Study and teaching–Methodology. 2. Science teachers–Training of. I. Title.

Q181.N78 2010
507.1–dc22

2010006479

ISBN13: 978-0-415-88237-8 (hbk)
ISBN13: 978-0-415-88238-5 (pbk)
ISBN13: 978-0-203-84753-4 (ebk)

10/17/11

To Magda and Sabuhi
For their love, patience and understanding

Contents

Preface ix
Acknowledgments xii

1 Introduction 1

2 The Role of Presuppositions, Contradictions, Controversies and Speculations versus Kuhn's "Normal Science" 17

3 A Rationale for Mixed Methods (Integrative) Research Programs in Education 34

4 Exploring Alternative Approaches to Methodology in Educational Research 49

5 Can Findings of Qualitative Research in Education be Generalized? 71

6 Qualitative Methodology and Its Pitfalls in Educational Research 86

7 Did Columbus Hypothesize or Predict? Facilitating Teachers' Understanding of Hypotheses and Predictions 103

8 Facilitating Teachers' Understanding of Alternative Interpretations of Conceptual Change 113

9 Progressive Transitions in Teachers' Understanding of Nature of Science Based on Historical Controversies 126

10 What "Ideas-About-Science" Should be Taught in School Science? 149

11 Whither Constructivism? Understanding the Tentative Nature of Scientific Knowledge 166

12 Conclusion: Methodologists Need to Catch Up with Practicing Researchers 187

Notes 199
References 200
Index 220

Preface

Research in science education has recognized the importance of history and philosophy of science (HPS). Over the last two decades there has been a worldwide sustained effort to introduce HPS in the science curriculum, textbooks and the classroom. Similarly, various reform efforts in different parts of the world have recognized the importance of presenting science to the students within an HPS perspective (e.g., *Project 2061* by the American Association for the Advancement of Science, AAAS). Implementation of these reform projects requires teacher training in order to facilitate an understanding of how science develops and the dynamics of scientific progress. Consequently, in order to change the educational landscape we need to familiarize teachers with developments in HPS so that they can teach science as practiced by scientists. Research has also shown that these aspects with respect to the nature of science have generally been ignored by textbooks, classroom teachers and some curriculum developers. This book provides a comprehensive overview of the contemporary history and philosophy of science and its implications for science teacher education.

History of science shows that most of the major achievements of what we now take as the advancement or progress of scientific knowledge have been controversial due to alternative interpretations of experimental data. Scientific controversies are found throughout the history of science. While nobody would deny that science in the making has had many controversies, most science textbooks and curricula consider it as the uncontroversial rational human endeavor.

This book is based on the following epistemological guidelines: (a) it is the problem to be researched that determines the methodology to be used; (b) a historical reconstruction of a scientific theory can determine the different sources that contributed to its development; and (c) discussion of the historical reconstructions based on interactions among classroom teachers can facilitate the elaboration of new teaching strategies. These guidelines have been followed in this book while discussing the different historical episodes, which have important implications for teacher training.

Based on these considerations my book presents an overview of the dynamics of scientific progress and then develops innovative teaching strategies based on actual classroom practice. Development of the teaching strategies in turn is anchored in high school and introductory level university teachers, who were participating in graduate courses. The sequence of courses (methodology, epistemology and research) was designed with the objective of progressively introducing various aspects of "science in the making". Classroom discussions were based on highlighting controversial aspects of various historical episodes included in the science curriculum. Participating teachers were not only exposed to the controversies but also encouraged to support, defend or critique the different interpretations. Just as the historical reconstructions discussed in class provide a glimpse of "science in the making", all chapters of this book facilitate an understanding of how teachers interact to critically appraise dynamics of scientific progress. Some of the salient features of my book are:

a. Historical reconstructions presented are very different from textbook presentations.
b. Historical and philosophical discussions are not simple adjuncts to the course but rather an essential part of the curriculum.
c. Science does not advance by just doing the experiments and having the data.
d. Progress in science inevitably leads to controversies and alternative interpretations of data.
e. Teachers' epistemological outlook is crucial in order to facilitate conceptual understanding.
f. Motivation of teachers to question the conventional wisdom with respect to progress in science (as depicted in textbooks) and pursue further studies within a history and philosophy of science perspective.
g. Given the opportunity, teachers can critically scrutinize the different historical episodes and suggest ways for innovating classroom practice.
h. Teaching science as practiced by scientists is an important guideline for teacher training.

In writing this book my objective was not any particular course. This has the advantage that the book could be adopted partially for various types of courses, such as: Introduction to history and philosophy of science; Research methodology; Dynamics of scientific progress; How to introduce nature of science in the classroom. My book explicitly deals with the following aspects: (a) teacher-training courses based on the experience of in-service teachers; (b) history and philosophy of science as an essential part of the science curriculum; (c) methodological (qualitative, quantitative, mixed methods, controversies, presuppositions, speculations, hypotheses, predictions); and (d) history and philosophy of science as part of classroom practice (alternative interpretations, nature of science, ideas about science, tentative nature of scientific knowledge). The intended audience for this book is: secondary and introductory level university teachers, science teacher educators, researchers in science education, science teachers, science methods course teachers and students and graduate students.

Chapters 2–11 of this book deal with different aspects of history and philosophy of science and how it can be incorporated in the classroom, and can easily constitute a course outline. Chapter 2 contrasts the role of presuppositions, contradictions, controversies and speculations (i.e., science in the making) with

Kuhn's "normal science". Based on this, Chapter 3 provides a rationale for mixed methods (integrative) research programs in education. Alternative approaches to methodology in educational research are explored in Chapter 4. Possibility of generalization in qualitative educational research is considered in Chapter 5. Difficulties associated with qualitative research in education is the subject of Chapter 6. Ability to formulate hypotheses and predictions is treated in Chapter 7. Alternative interpretations of conceptual change based on rival theories are discussed in Chapter 8. Role of historical controversies and their application in the classroom is the subject of Chapter 9. Chapter 10 considers which ideas about science should be included in the classroom based on a historical perspective. Finally, based on constructivism, understanding tentative nature of scientific knowledge is illustrated in Chapter 11. Contents of this book can be divided into three main groups: (a) Chapters 2 and 3 primarily deal with philosophical questions; (b) Chapters 4–7 are based on methodological problems; and (c) Chapters 8–11 illustrate how history and philosophy of science can be introduced in the classroom. Based on their interests and orientation readers can select the appropriate chapters.

Acknowledgments

This book has been in preparation for almost 20 years, in which I have interacted and received feedback from many colleagues, friends and my students. Looking back over these years, I had no idea that this work would take the form of a book. My institution, Universidad de Oriente (Venezuela), has supported most of my research activities. Juan Pascual-Leone (York University, Toronto) has been a major source of inspiration for understanding cognitive psychology and later my interest in history and philosophy of science. I have benefited immensely from discussions and criticisms at different stages from: Richard F. Kitchener (Colorado State University), Art Stinner (University of Manitoba), Stephen Klassen (University of Winnipeg), Michael R. Matthews (University of New South Wales), Stephen G. Brush (University of Maryland) and Gerald Holton (Harvard University).

I would like to thank the three reviewers who provided constructive criticisms and at the same time encouragement for completing the book. Louis Rosenblatt provided insight with respect to the tentative nature of science in situations where different interpretations are offered and views held despite seeming refutation. William Cobern (Western Michigan University) pointed out the folly of ideological decisions on research methods, namely the research questions need to drive our research methods. Chin-Chung Tsai (National Taiwan University of Science and Technology) considered the analogies between physical science experiments and social science research to be helpful for educational research.

A special word of thanks is due to Naomi Silverman, Senior Editor at Routledge (New York) for her enthusiastic support throughout the different stages of preparing the manuscript and publication.

Thanks are due to the following publishers for reproduction of materials from my publications: Elsevier (Chapters 2 and 11); Wiley-Blackwell (Chapter 3); Taylor & Francis (Chapter 7); and Springer (Chapters 4, 5, 6, 8, 9 and 10).

Chapter 1

Introduction

Most science teachers, textbooks and curricula consider progress in science to be based entirely on experiments, which provide evidence that unambiguously leads to the formulation of scientific theories. A historical reconstruction of the different topics of the science curriculum reveals that although experiments are important, interpretation of the data is even more important. In order to develop their research programs, besides the experimental data, scientists rely on their guiding assumptions (presuppositions), which inevitably leads to conflicts and controversies. Review of the literature based on textbook analyses reveals almost a complete lack of understanding of the role played by presuppositions, contradictions, controversies and speculations (Niaz, 2008a). In the early stages of all research, scientists are groping with difficulties, future of the research cannot be predicted, interpretations are uncertain and stakes are high due to competing groups (peer pressure). Furthermore, students' understanding of nature of science is quite similar to that of the textbook. The traditional science curriculum in general would seem to ignore the "how" and "why" of science in the making. Studies presented in this book suggest that the teacher, by "unfolding" the different episodes (based on historical reconstructions), can emphasize and illustrate how science actually works, namely tentative, controversial rivalries among peers and alternative interpretations of data. Consequently, innovating science teacher education is an important part of the research agenda.

According to Gage (2009), as compared to other areas in education, research on teaching has been neglected and suggests the following topics for research: need for a theory, evolution of a paradigm for the study of teaching, conception of the process of teaching, conception of the content of teaching, conception of students' cognitive capabilities and motivations, conception of classroom management and the integration of these conceptions. Borko, Liston and Whitcomb (2007) have also recognized that teacher education is relatively a new field of study. Furthermore, these authors have emphasized the importance of research in teacher education and suggested:

> Several sound research genres are available to the teacher education research community, each genre better suited for some questions than others. *The researcher's first and most essential role is to pose questions of practical and*

> *theoretical significance*. Researchers then should evaluate which genre or combination of genres best fits the question(s) and the resources available to conduct a well-designed study.
>
> (p. 9, emphasis added)

This is sound advice, in view of the fact that most methodology courses suggest that researchers should first select the genre of research (qualitative, quantitative, mixed, etc.) and then the question to be investigated. A leitmotiv of this book is that it is the problem to be researched that determines the methodology to be used. It seems that after the paradigm wars (Gage, 1989; Phillips, 1983), the research community has learned that we cannot adopt the research methodology a priori but rather let the problem situation provide the rationale and guidelines. This is a major step in going beyond the paradigm wars (Saloman, 1991).

The Arizona Collaborative for Excellence in the Preparation of Teachers (ACEPT) Program is one of several reform efforts supported by the National Science Foundation in the USA. The primary ACEPT reform mechanism has been month-long summer workshops in which university and community college science and mathematics faculty learn about instructional reforms and then attempt to apply them in their courses. Adamson et al. (2003) studied whether enrollment of pre-service teachers in one or more of these ACEPT-reformed undergraduate courses is linked to the way they teach after they graduate and become in-service teachers and concluded: "These results support the hypothesis that teachers teach as they have been taught. Furthermore, it appears that instructional reform in teacher preparation programs including both methods and major's courses can improve secondary school student achievement" (pp. 939–940). If "teachers teach as they have been taught" then innovating teacher training programs is all the more important. Teachers not only contribute to the development of individuals and societies but also attain self-realization through teaching (Shim, 2008).

In a recent survey conducted among members of the National Association for Research in Science Teaching (NARST) to determine the importance of issues faced by the science education community, the two top priorities were enhancing in-service teacher education and improving pre-service teacher education (cf. Czerniak, 2009). Given the presence of NARST members both in the USA and many other countries, it seems that teacher training constitutes an important part of the science education research agenda.

Historians and philosophers of science have devoted a considerable amount of work toward understanding the dynamics of scientific progress and what constitutes nature of science, NOS (Giere, 2006; Niaz, 2009a). In contrast, most students and teachers in most parts of the world frequently believe that science is a collection of facts and that the best way to learn science is to memorize those facts (Linn, Songer & Lewis, 1991). Millar (1989) has cautioned against perceiving nature of science as an empiricist epistemology, for the following reasons: (a) pedagogical: teaching science becomes a business of rote memorization of standard facts, laws, theories, methods and problem-solving procedures; and (b) epistemological: science is viewed as infallible and a body of absolute facts or received knowledge. The degree to which students' conceptions of NOS are influenced by their teachers and textbooks is the subject of considerable research. According to Lederman (1992), such influence is mediated by a complex set of factors, such as curriculum constraints, administrative policies and teachers' conceptualization of learning. Given the complexity and multifaceted nature of the issues involved and a running controversy among philosophers of science themselves, implementation of NOS in the classroom has also been difficult. Despite the controversy a certain degree of consensus has been achieved within the science education community and nature of science can be characterized, among others, by the following aspects (Abd-El-Khalick, 2004; Lederman, 2004; McComas et al., 1998; Niaz, 2001a, 2008b; Osborne et al., 2003; Scharmann & Smith, 2001; Smith & Scharmann, 1999):

1. Scientific knowledge relies heavily, but not entirely, on observation, experimental evidence, rational arguments and skepticism.
2. Observations are theory-laden.
3. Science is tentative/fallible.
4. There is no one way to do science and hence no universal, recipe-like, step-by-step scientific method can be followed.
5. Laws and theories serve different roles in science and hence theories do not become laws even with additional evidence.
6. Scientific progress is characterized by competition among rival theories.
7. Different scientists can interpret the same experimental data in more than one way.
8. Development of scientific theories at times is based on inconsistent foundations.
9. Scientists require accurate record keeping, peer review and replicability.
10. Scientists are creative and often resort to imagination and speculation.
11. Scientific ideas are affected by their social and historical milieu.

A review of the literature shows that most teachers in many parts of the world lack an adequate understanding of some or all of the different NOS aspects outlined above (Akerson et al., 2006; Bell et al., 2001; Blanco & Niaz, 1997; Clough, 2006; Dogan & Abd-El-Khalick, 2008; Lederman, 1992; Mellado et al., 2006; Pomeroy, 1993; Tsai, 2002). This should be no surprise to anyone who has analyzed science curricula and textbooks, which have a pronounced stance toward an entirely empiricist and positivist epistemology. Tsai (2006) has argued cogently for including the various aspects of NOS for both pre-service and in-service teacher training:

> Scientific knowledge should be regarded as an invented reality, which is also constructed through the use of agreed-upon paradigms, acceptable form of evidence, social negotiations in reaching conclusions, and technological, contextual and cultural impacts are recognized by participating scientists. These views are very different from traditionally *empiricist* perspectives. The empiricist position assumes that scientific knowledge is a discovery of an

> objective reality external to ourselves and discovered by observing, experimenting or application of a universal scientific method.
>
> (pp. 363–364, original italics, underline added)

Let us now compare this with what Steven Weinberg (2001), Nobel Laureate in physics, has to say about objective reality and truth in science: "What drives us onward in the work of science is precisely the sense that there are *truths out there to be discovered*, truths that once discovered will form a permanent part of human knowledge" (p. 126, emphasis added). No wonder science curricula and textbooks in most parts of the world follow a similar epistemology. Giere (2006) has characterized such philosophical positions as "objectivist realism" (p. 5), and explained cogently:

> Weinberg should not need reminding that, at the end of the nineteenth century, physicists were as justified as they could possibly be in thinking that classical mechanics was objectively true. That confidence was shattered by the eventual success of relativity theory and quantum mechanics a generation later.
>
> (p. 118)

This leads to yet another interesting issue: do all Nobel Laureates in physics follow "objectivist realism"? The following statement from Leon Cooper, another Nobel Laureate in physics, can provide science teachers a better insight with respect to the dynamics of scientific progress:

> Observations can have varying interpretations, but this does not undermine the objective nature of science … It's somewhat ironic that what we like to call the meaning of a theory, its interpretation, is what changes. Think, for example, of the very different views of the world provided by quantum theory, general relativity and Newtonian theory.
>
> (p. 47, reproduced in Niaz, Klassen, McMillan & Metz, 2010a)

As a methodological guideline (important for teacher training), Giere (2006) suggests that only a historical examination of a scientific theory can determine the different sources that contributed to its development (p. 6). Similarly, Phillips (2005a) has critiqued educational research for not providing real examples and concluded that philosophy of educational research is roughly at the stage that much philosophy of science was six decades ago (Phillips is referring to the in-depth historical studies starting in the 1950s by contemporary philosophers of science, such as Popper, Kuhn, Lakatos, Cartwright and Galison). In contrast to Giere's (2006) "objectivist realism", Cobern and Loving (2008) have espoused an "epistemological realism" with the following caveat, "science is imperfect, incomplete and fallible; and is not the only source of knowledge that we as humans find of value" (p. 443). These critiques and reflections have served as a guideline in the elaboration of the different historical episodes in this book (especially Chapter 3) and their implications for teacher training.

Kenneth Wilson, another Nobel Laureate in physics, has argued forcefully as to how the "perpetual flux" in the history of science can cultivate students' expectations of how they might contribute to future changes in scientific innovation:

> The key role of history here is characterizing the complexities of how science *changes.* So many science textbooks unhelpfully—and above all inaccurately—cultivate a rather static image of scientific disciplines, as if they were completed with comprehensive certainty. It is perhaps not difficult to understand how this gross oversimplification might arise as the result of a pedagogical need to "tidy up" the presentation of science to meet the needs and capacities of students. But faced with the textbook spectacle of such an apparently unalterable monolith, is it any wonder that students can have difficulty conceiving how they might ever contribute to science?
>
> (Gooday, Lynch, Wilson & Barsky, 2008, p. 326, original italics)

Wilson and Barsky (1998) have provided the lead in integrated historical teaching in order to enable students to understand what science is and how it is conducted. They have suggested that in order for these reform efforts to be successful, teacher preparation is a critical issue.

Slater (2008) has raised a provocative question for science teacher education: how to justify teaching false science? This, in turn, is based on the premise that we teach false science (e.g., Newtonian mechanics, Thomson, Rutherford and Bohr models of the atom). As a possible solution to the dilemma, Slater suggests that "the best way of teaching false science is by teaching it *as false*, but *illustratively*—incorporating a critical historical perspective into the science curriculum" (p. 541, original italics). This clearly shows the need for incorporating a history and philosophy of science perspective in the science curriculum, in order to facilitate a better understanding of the dynamics of scientific progress. In other words, "false science" can illustrate to students and teachers how understanding of experimental data in the history of science led to controversies and alternative interpretations.

With this background it would help to better understand the considerable amount of work that has been done to teach NOS in the classroom (Abd-El-Khalick & Akerson, 2004, 2007; Bianchini & Colburn, 2000; Ford & Wargo, 2007; Irwin, 2000; Khishfe & Lederman, 2006; Lin & Chen, 2002; Niaz et al., 2002; Southerland et al., 2006; Sowell et al., 2007; Von Aufschnaiter et al., 2008; Wong et al., 2008). Nevertheless, the relationship between teachers' conceptions of NOS and their classroom practice is more complex than generally appreciated. Abd-El-Khalick and Lederman (2000b) have attributed this to various factors, such as: pressure to cover content, classroom management and organizational principles, concern for student abilities and motivation, institutional constraints, teaching experience and difficulties in understanding the philosophical underpinnings of nature of science. Concern for covering content is counterproductive if we want to cultivate students' interest and motivation with respect to what is science and how it progresses, and at the same time foster a natural curiosity about the world around us. Cobern et al. (1999) have argued cogently with respect to how students' understanding of nature of science can be "successful only to the extent that science finds a niche in the cognitive and cultural milieu of students" (p. 541).

Despite the difficulties, research in science education has continued to work on the development and implementation of courses/materials both at the undergraduate and high school levels, in order to facilitate students' and teachers' understanding of NOS (Abd-El-Khalick, 2005; Abd-El-Khalick, Bell & Lederman, 1998; Niaz, 2009b; Pocoví, 2007). At this stage it would be interesting to provide greater insight into how teachers can acquire a deeper understanding of the nature of science and how progress in science is a complex process. Sadler et al. (2004) have argued cogently that science operates under the implicit assumption that scientific knowledge develops, builds upon itself and changes over time, namely, its tentative nature. Furthermore, scientists would not devote their lives to the pursuit of knowledge if they had no chance of adding to or changing prevailing paradigms. One of the participating teachers in a study designed to facilitate greater understanding of NOS provided the following informed view with respect to how observations are theory-laden:

> Science is not as objective as people would like to believe. When presented with evidence, people interpret it differently. The scientists involved in the debate about extinction of dinosaurs each came from different paradigms. They interpret their evidence according to their own paradigm.
>
> (Reproduced in Abd-El-Khalick, 2005, p. 29)

A critical reader may point out that such thinking may lead the teachers to consider decisions in the construction of scientific knowledge as arbitrary. However, this is not the intention. The important point is to understand that objectivity by itself does not help to take decisions, but rather it is the decision-making process (controversy, conflicts and alternative interpretations of data) that provides an objective status to the scientific enterprise. Campbell (1988a), a methodologist, has expressed this in succinct terms:

> [T]he objectivity of physical science does *not* come from the fact that single experiments are done by reputable scientists according to scientific standards. It comes instead from a social process which can be called competitive cross-validation ... and from the fact that there are many independent decision makers capable of rerunning an experiment, at least in a theoretically essential form. The resulting dependability of reports ... comes from a social process rather than from dependence upon the honesty and competence of any single experimenter.
>
> (pp. 302–303, original italics)

A major difficulty in implementing NOS is the expectation that students will come to understand it by "doing science" (Lederman, 2004, p. 315). This is like assuming that students would come to understand photosynthesis just by watching a plant grow. In order to facilitate understanding of NOS teachers need to go beyond the traditional curriculum and emphasize the difficulties faced by the scientists and how interpretation of data is always problematic, leading to controversies among contending groups of researchers. Next, examples are provided of how "doing science" is not a sufficient condition for understanding science.

J.J. Thomson (1897) is generally credited to have "discovered" the electron while doing experiments with cathode rays. Determination of the mass-to-charge (*m/e*) ratio of the cathode rays can be considered the most important experimental contribution of Thomson. Yet, he was neither the first to do so nor the only experimental physicist. Kaufmann and Wiechert also determined the *m/e* of cathode rays in the same year and their values agreed with each other (for details, see Niaz, 1998). If we demonstrate this experiment in the classroom or students handle the equipment themselves (i.e., doing science), it may be useful, and this is good educational practice. However, by emphasizing that "science is empirical" (doing experiments) we shall be denying students an important aspect of the nature of science, namely what made Thomson's work different from that of Kaufmann and Wiechert. Falconer (1987) has explained cogently how both Kaufmann and Wiechert lacked a theoretical framework (heuristic principle) to understand the data. In contrast, Thomson had a heuristic principle before doing the experiments, namely cathode rays could be considered as ions (if *m/e* ratio was not constant) or universal charged particles (if *m/e* ratio was constant). Indeed, most general chemistry and physics textbooks emphasize the experimental details (doing science) and ignore Thomson's heuristic principle for interpreting and understanding the data (for details, see Niaz, 1998; Rodríguez & Niaz, 2004a).

Soon after Geiger and Marsden (1909) published their results (working under E. Rutherford's supervision), Thomson and colleagues also started working on the scattering of alpha particles in their laboratory (again, doing the experiment in the classroom can help). Although experimental data from both laboratories were similar, interpretations of Thomson and Rutherford were entirely different. Thomson propounded the hypothesis of *compound scattering*, according to which a large-angle deflection of an alpha particle resulted from successive collisions between the alpha particles and the positive charges distributed throughout the atom. Rutherford (1911), in contrast, propounded the hypothesis of *single scattering*, according to which a large-angle deflection resulted from a single collision between the alpha particle and the massive positive charge in the nucleus. The rivalry led to a bitter dispute between the proponents of the two hypotheses (for details, see Niaz, 1998; Wilson, 1983). At one stage the controversy became so bitter that Rutherford charged that a colleague of Thomson had "fudged" the data. Once again, most chemistry and physics textbooks ignore the difficulties involved in understanding the data and the ensuing controversy (cf. Niaz, 1998; Rodríguez & Niaz, 2004a).

History of science shows how R.A. Millikan (1868–1953) and F. Ehrenhaft (1879–1952) obtained very similar experimental observations (oil drop experiment), and yet their theoretical frameworks led them to postulate the elementary electrical charge (electrons) and fractional charges (sub-electrons), respectively. The Millikan–Ehrenhaft controversy lasted for many years (1910–1923) and was discussed by leading scientists. The problematic nature of Millikan's interpretation was revealed many years later when Holton (1978a, 1978b) consulted his handwritten notebooks in CALTECH. The oil drop experiment is still used in undergraduate physics labs and continues to be problematic for students (cf. Klassen, 2009). Not surprisingly, both general chemistry and physics textbooks do present the experiment in considerable detail, and still completely ignore the Millikan–Ehrenhaft controversy (Niaz, 2000a; Rodríguez & Niaz, 2004b).

Experiments related to the photoelectric effect played a crucial role in the construction of the modern atomic theory and form an important part of the science curriculum. Once again, Robert Millikan provided the first experimental evidence for Einstein's photoelectric equation. Interestingly, however, in the same publication (Millikan, 1916), he recognized the validity of Einstein's equation and simultaneously questioned the underlying hypothesis of lightquanta put forward by Einstein. This may sound incredible to any student who has not been exposed to history and philosophy of science. Philosophers of science refer to this as underdetermination of scientific theories by experimental evidence, namely no amount of experimental evidence can provide conclusive proof for a theory (for details, cf. Niaz, 2009a). A recent study has revealed an almost complete lack of the historical perspective (essential for conceptual understanding) in presenting the photoelectric effect in general physics textbooks (cf. Niaz, Klassen, McMillan & Metz, 2010b). These authors reported that a great majority of the textbooks considered that Millikan had provided experimental evidence for Einstein's hypothesis of lightquanta, contrary to what he himself had claimed.

These examples provide a clear illustration of the dilemma involved in "doing science" and understanding science, as teachers in most parts of the world invariably emphasize the former, that is, lab activities, and thus do not arouse students' curiosity with respect to "science in the making". Interestingly, Tsai (2003) has investigated laboratory learning environments and found that teachers generally held an empiricist epistemology and showed higher preferences for better equipment than did their students. Cathode ray experiments, scattering of alpha particle experiments, photoelectric effect and the oil drop experiments are considered to be the foundation of modern science (early 20th century) and are included in science curricula and textbooks both at the upper secondary and university freshman level, in almost all parts of the world (Chapters 2 and 3 provide more details of these and other experiments). However, very rarely are students provided an insight into what the scientists were discussing/arguing with their peers while the experiments were being conducted. In other words, scientific theories require a considerable amount of ingenuity, creativity and "competitive cross-validation" in order to convince the scientific community. A major objective of this book is to provide guidelines and a framework for including these historical episodes in the upper secondary and university freshman classroom practice (see Chapters 8, 9, 10 and 11). In order to facilitate understanding, a brief overview of the different chapters of this book is presented next.

Role of presuppositions, contradictions, controversies and speculations versus Kuhn's normal science. Kuhn (1970) considered textbooks to be good "pedagogical vehicles" for the perpetuation of "normal science" (Chapter 2).

Collins (2000) has pointed out a fundamental contradiction with respect to what science could achieve (discover and create new knowledge) and how we teach science (dogmatic and authoritarian). Despite the reform efforts (*Project 2061, Beyond 2000*), students (secondary and university) still have naive views about the nature of science in which experimental data unambiguously lead to the formulation of laws and theories. Review of the literature based on textbook analyses shows an almost complete lack of understanding of the role played by presuppositions, contradictions, controversies and speculations in scientific progress. Kuhn's advice based on "normal science" would seem to suggest that the science curriculum need not appeal to the imagination and creativity of the students. It is not my intention to suggest that Kuhn has promoted the inclusion of "normal science" in science textbooks. The teacher by "unfolding" the different episodes (based on historical reconstructions) can emphasize and illustrate how science actually works (tentative, controversial, rivalries, alternative interpretations of the same data), and this will show to the students that they need to go beyond "normal science" as presented in their textbooks.

A rationale for mixed methods (integrative) research programs in education. Recent research shows that research programs (quantitative, qualitative and mixed) in education are not displaced (as suggested by Kuhn) but rather lead to integration. The objective of Chapter 3 is to present a rationale for mixed methods (integrative) research programs based on contemporary philosophy of science (Lakatos, Giere, Cartwright, Holton, Laudan). This historical reconstruction of episodes from physical science (spanning a period of almost 300 years, from the 17th to the 20th century) does not agree with the positivist image of science. Quantitative data (empirical evidence), by itself, does not facilitate progress (despite widespread belief to the contrary), neither in the physical sciences nor in the social sciences (education). A historical reconstruction shows that both Piaget and Pascual-Leone's research programs in cognitive psychology follow the Galilean idealization quite closely, similar to the research programs of Newton, Mendeleev, Einstein, Thomson, Rutherford, Millikan and Perl in the physical sciences. This relationship does not imply that researchers in education have to emulate research in the physical sciences. A major argument in favor of mixed methods (integrative) research programs is that it provides a rationale for hypotheses, theories, guiding assumptions and presuppositions to compete and provide alternatives. Similar to the physical sciences, this proliferation of hypotheses leads to controversies and rivalries, and thus facilitates the decision-making process of the scientific community.

Exploring alternative approaches to methodology in educational research. The objective of Chapter 4 is to provide in-service teachers an opportunity to familiarize themselves with the controversial nature of progress in science (growth of knowledge) and its implications for research methodology in education. The study is based on 41 participants who had registered for a 9-week course on Methodology of Investigation in Education, as part of their Master's degree program. The course is based on 20 readings drawing on a history and philosophy of science perspective (positivism, constructivism, Popper, Kuhn, Lakatos) and its implications for educational research (Campbell, Erickson). Course activities included written reports, classroom discussions based on participants' presentations and written exams.

Can findings of qualitative research in education be generalized? Most qualitative researchers do not recommend generalization from qualitative studies, as this research is not based on random samples and statistical controls. The objective of Chapter 5 is to explore the degree to which in-service teachers understand the controversial aspects of generalization in both qualitative and quantitative educational research and as to how this can facilitate problems faced by the teachers in the classroom. The study is based on 83 participants who had registered for a 10-week course on Methodology of Investigation in Education, as part of their Master's degree program. The course is based on 11 readings drawing on a philosophy of science perspective (positivism, constructivism, Popper, Kuhn, Lakatos). Course activities included written reports, classroom discussions based on participants' presentations and written exams.

Qualitative methodology and its pitfalls in educational research. There is considerable controversy in educational research with respect to the use of qualitative and quantitative data and as to what constitutes scientific research. The objective of Chapter 6 is to explore the degree to which in-service teachers understand the difference between qualitative/quantitative data and methods, validity/authenticity, generalization and how these can be used to solve problems faced by the teachers. The study is based on 84 participants who had registered for a 10-week course on Methodology of Investigation in Education, as part of their Master's degree program. The course is based on 11 readings drawing on a history and philosophy of science perspective (positivism, constructivism, Popper, Kuhn, Lakatos). Course activities included written reports, classroom discussions based on participants' presentations and written exams.

Did Columbus hypothesize or predict? Facilitating teachers' understanding of hypotheses and predictions. A review of the literature in science education shows that most students have difficulties in hypothetico-deductive reasoning. The ability to elaborate and differentiate between observations, hypotheses and predictions is important and need not necessarily be considered as part of the scientific method. Most philosophers of science would question the existence of a scientific method as a series of specifiable procedures that constitute an algorithm (Cartwright, 1999; Giere, 1999; Lakatos, 1970; Polanyi, 1964). The objective of Chapter 7 is to investigate high school and freshman university teachers' ability to understand the difference between *hypotheses* and *predictions* in the everyday context of Columbus' discovery of America. Eighty-three high school and introductory level university teachers enrolled in a Methodology course were asked to elaborate and explain a prediction and a hypothesis based on Columbus' discovery. As a follow up, a study was designed to facilitate in-service high school and university teachers' understanding of the difference between the terms *hypothesis* and *prediction*. The context for understanding these terms was Columbus' discovery of America (same as in the previous study). Control-group teachers (n=94) were evaluated before the discussion of these terms, whereas Experimental group teachers (n=102) were evaluated after these terms had been fully discussed and elaborated in class.

Facilitating teachers' understanding of alternative interpretations of conceptual change. Historians and philosophers of science have recognized the importance of controversies in the progress of science. The objective of Chapter 8 is to facilitate in-service chemistry teachers' understanding of conceptual change based on alternative philosophical interpretations (controversies). Selected controversies formed part of the chemistry curriculum both at secondary and university freshman level. The study is based on 17 in-service teachers who had registered for an 11-week course on Investigation in the Teaching of Chemistry as part of their Master's degree program. The course is based on 17 readings drawing on a history and philosophy of science perspective with special reference to controversial episodes. Course activities included written reports, classroom discussions based on participants' presentations and written exams. In this study most of the teachers went through an experience that involved inconsistencies, conflicts, contradictions and finally some degree of conceptual change. A few of the participants, however, resisted any change, but still raised important issues with respect to conceptual change.

Progressive transitions in teachers' understanding of nature of science based on historical controversies. The objective of Chapter 9 is to facilitate progressive transitions in chemistry teachers' understanding of NOS in the context of historical controversies. Selected controversies referred to episodes that form part of the chemistry curriculum both at secondary and university freshman level. The study is based on 17 in-service teachers who had registered for an 11-week course on Investigation in the Teaching of Chemistry as part of their Master's degree program. The course is based on 17 readings drawing on a history and philosophy of science perspective with special reference to controversial episodes in the chemistry curriculum. Course activities included written reports, classroom discussions based on participants' presentations and written exams. The opportunity to reflect, discuss and participate in a series of course activities based on controversies can enhance teachers' understanding of NOS.

What "ideas-about-science" should be taught in school science? The objective of Chapter 10 is to facilitate in-service chemistry teachers' understanding of nature of science and what "ideas-about-science" can be included in the classroom. The study is based on 17 in-service teachers who had registered for an 11-week course on Epistemology of Science Teaching as part of their Master's degree program. The course is based on 17 readings drawing on NOS and its critical evaluation. Course activities included written reports, classroom discussions based on participants' presentations and written exams. This course provided participant teachers an opportunity to familiarize themselves with research on what "ideas-about-science" can be taught in the classroom and how critical appraisal of the literature is necessary in order to go beyond our present understanding of the issues.

Whither constructivism? Understanding the tentative nature of scientific knowledge. Constructivism in science education has been the subject of considerable debate in the science education literature. The purpose of Chapter 11 is to facilitate chemistry teachers' understanding that the tentative nature of scientific knowledge leads to the coexistence of rivalries among different forms of constructivism in science education. The study is based on 17 in-service teachers who had registered for an 11-week course on Epistemology of Science Teaching as part of their Master's degree program. The course is based on 17 readings drawing on NOS and a critical evaluation of constructivism. Course activities included written reports, classroom discussions based on participants' presentations and written exams.

At this stage I would like to introduce some basic ideas that may be of help, especially to students who may not be familiar with recent developments in history and philosophy of science.

Positivism

It would be helpful to have a historical perspective with respect to the various forms of positivism (Phillips, 1994a). History of science shows that positivism was the dominant philosophy from about the end of the 19th century to about the middle of the 20th century. Positivism has many faces and philosophers tend to characterize it in different ways: (a) *classic positivism* can be traced to Comte (1798–1857), who emphasized that science focuses upon observation and hence scientific knowledge consisted only in the description of observed phenomena and not inferred theoretical entities; (b) *logical positivism* associated with the Vienna Circle which was very active during the 1930s and introduced the Verifiability Principle, according to which something is meaningful if and only if it is verifiable empirically, or, in other words, "if it can't be seen or measured, it is not meaningful to talk about"; (c) *behaviorism* for their hostility to abstract theorizing and metaphysics; and (d) *empiricism* which again emphasizes that our knowledge is wholly or partly based on experience through the senses and introspection. According to Phillips (1983), although logical positivism is a type of empiricism, not all varieties of empiricism are positivistic.

The importance of having positivist or more adequate epistemological views is important for teacher training. For example, Tsai (2007) has explored the relationship between middle school physical science teachers' (Taiwan) epistemological views, teaching beliefs, instructional practices and students' epistemological views. Findings suggested adequate coherence between teachers' epistemological views and teaching beliefs as well as instructional practices. Teachers with relatively positivist-aligned views tended to draw attention to students' science scores in tests and allocate more instructional time on teacher-directed lectures and in-class examinations, thus implying more passive or rote learning. In contrast, teachers with constructivist-oriented views tended to focus on student understanding and application of scientific concepts, by devoting more time to inquiry activities and interactive discussions. This clearly shows that teachers with positivist views tend to encourage and foster more traditional teaching practices based on algorithmic learning.

Similarly, logical positivism has also been the subject of study with special reference to the science curriculum. According to Van Aalsvoort (2004), most secondary school students consider chemistry to be irrelevant. Based on a review of science education literature, the atheoretical nature of the observational language and the curriculum (based heavily on the textbooks), the author concluded that chemical education is driven by logical positivism. As a philosophy of science, logical positivism creates a divide between science and society. Based on these premises, the author hypothesized that the adoption of logical positivism causes chemistry's lack of relevance in chemical education. This hypothesis was substantiated by an analysis of the secondary school chemistry curriculum in the Netherlands. Based on these considerations, the author concluded,

> Chemical education is relevant from a social point of view to the extent that the knowledge it provides is applicable to solve society's problems. Yet, due to the hierarchical relation between scientific knowledge and its applications, the former is preferred above the latter in chemical education, thereby leaving the relevance of chemical knowledge for society mostly out of sight. (p. 1166)

Finally, the author suggested that as an alternative to logical positivism, science educators could explore activity theory within the sociocultural approach.

Galilean Idealization

In contrast to Aristotle, who believed that a continually acting cause (i.e., force) was necessary to keep a body moving horizontally at a uniform velocity, Galileo predicted that if a perfectly round and smooth ball was rolled along a perfectly smooth horizontal endless plane there would be nothing to stop the ball (assuming no air resistance), and so it would roll on forever. Galileo, however, did not have the means to demonstrate that Aristotle was wrong, so he asked an epistemological question: what would make it (body) stop? And then went on to argue that under ideal conditions (with impediments, such as shape of the ball and the surface, controlled) a ball could roll on forever. Similarly, Galileo's discovery of the law of free fall later led to a general constructive model of falling bodies (Pascual-Leone, 1978). The law in its modern form can be represented by: $s = 1/2\ g\ t^2$ (s = distance, t = time and g = a constant). In order to "prove" his law of free fall, Galileo should have presented empirical evidence to his contemporaries by demonstrating that bodies of different weight (but of the same material) fall at the same rate. If the leaning tower of Pisa mythical experiment (cf. Segre, 1989, for recent controversy) was ever conducted, it would have shown Galileo to be wrong. According to Pascual-Leone (1978), empirical computation of the value of s as a function of the variable t, "where vacuum and other *simplifying assumptions* are not satisfied" (emphasis added, p. 28), would lead to a rejection of the law. As a direct empirical test of Galileo's ideal law was not possible, he used his inclined plane experiment to show that as the angle of incidence approximated 90° (free fall), the acceleration of objects rolling down an inclined plane increasingly approximated a constant. According to Kitchener (1993, p. 142), by extrapolation one may assume it is also true of free fall as a limiting case.

Following Galileo's method of idealization (considered to be at the heart of all modern physics by Cartwright, 1989, p. 188) scientific laws, being epistemological constructions, do not describe the behavior of actual bodies. According to Lewin (1935), for example, the law of falling bodies refers only to cases that are never realized, or only approximately realized. Only in experiment, which is under artificially constructed conditions (idealization), do cases occur which approximate the event with which the law is concerned. Furthermore, Lewin has argued that this conflict between quantification (Aristotelian) and qualitative understanding (Galilean) modes of thought constitutes a paradox of empiricism. Galileo's law of free fall, Newton's laws, gas laws they all describe the behavior of ideal bodies that are abstractions from the evidence of experience and the laws are true only when a considerable number of disturbing factors, itemized in the *ceteris paribus* clauses, are eliminated (cf. Ellis, 1991; Matthews, 1987; McMullin, 1985; Niaz, 1999a). *Ceteris paribus* clauses play an important role in scientific progress, enabling us to solve complex problems by introducing simplifying assumptions (idealization). Lakatos (1970) has endorsed this position in the following terms: "Moreover, *one can easily argue that ceteris paribus clauses are not exceptions, but the rule in science*" (p. 102, original italics). This illustrates quite cogently the research methodology of idealization utilized for studying physical laws in particular and complex problems in general.

McMullin (1985) considers the manipulation of variables (disturbing factors) as an important characteristic of Galilean idealization:

> The move from the complexity of nature to the specially contrived order of the experiment is a form of idealization. The diversity of causes found in Nature is reduced and made manageable. The influence of impediments, i.e., causal factors which affect the process under study in ways not at present of interest, is eliminated or lessened sufficiently that it may be ignored.
>
> (p. 265)

According to Rigden and Stuewer (2005), in the physical sciences, the quantitative stands in sharp contrast to the qualitative. To understand any substantive topic, qualitative understanding is important, which requires a process of internalization so that an individual can draw on his resource of words to embrace a subject meaningfully. Further details are provided by Niaz (2005a).

Kuhn's Paradigms

According to Kuhn (1970), most scientific work consists of routine resolution of problems, which constitutes "normal science". As scientists working in a field of research achieve consensus with respect to a certain theoretical framework, it leads to the formation of a paradigm, which Kuhn later referred to as a "disciplinary matrix". While solving routine problems, scientists come up with anomalies that are difficult to resolve and the accumulation of such anomalies frequently leads to the overthrow of the existing paradigm and the revolutionary period that ensues leads to the formation of a new paradigm. Kuhnian philosophy of science has been a major source of inspiration for science educators and the following aspects of his philosophy have played an important role: (a) it presupposes subjectivity as an integral part of the scientific process, once thought to be wholly objective; (b) it asserts that different paradigms are incommensurate because their core beliefs are resistant to change and hence do not permit dialogue; (c) paradigms do not merge over time, rather they displace each other after periods of chaotic upheaval or scientific revolution; (d) Kuhnian displacements are not subtle events, but are rather understood as cataclysmic clashes in which losers languish and victors flourish. Kuhn (1970) himself referred to the subject in the following terms:

> if I am right that each scientific revolution alters the historical perspective of the community that experiences it, then that change of perspective should affect the structure of postrevolutionary textbooks and research publications. One such effect a shift in the distribution of the technical literature cited in the footnotes to research reports ought to be studied as a possible index to the occurrence of revolutions.
>
> (p. ix)

Lakatos' Research Programs

In contrast to paradigms (Kuhn), Lakatos (1970) postulates the importance of research programs that are formed by the *hard-core/negative heuristic* and the *positive heuristic*. Negative heuristic is based on the theoretical framework (presuppositions) of the scientist and is not necessarily refuted by experimental evidence. Most scientists before entering the laboratory do have their presuppositions and they hope to get experimental evidence for corroboration. The positive heuristic, on the other hand, defines problems, outlines the construction of a "protective belt" of auxiliary hypotheses, foresees anomalies and suggests solutions. Auxiliary hypotheses, for example, help the scientist to protect the hard-core of their research programs. An important aspect of the Lakatos methodology is to evaluate rival research programs on a continuum between progressive and degenerate. A research program is said to be *progressing* as long as its theoretical growth anticipates its empirical growth, that is, as long as it keeps predicting novel facts with some success that is "progressive problemshifts" (Lakatos, 1971, p. 100). A research program is progressing if it frequently succeeds in converting anomalies into successes, that is, explainable by the theory. The classic example of a successful research program is Newton's gravitational theory. The negative heuristic in Newton's program is the law of gravitation and his three laws of dynamics. The positive heuristic enables the scientist to build models by ignoring the *actual* counterexamples, the available data (Lakatos, 1970, p. 135).

Application of the Lakatosian methodology to Bohr's research program as an example of how scientists progress from simple to complex models (simplifying assumptions) is quite instructive. Lakatos (1970) differentiates clearly between the negative and positive heuristic of Bohr's research program. Bohr's (1913) famous four postulates constituted the negative heuristic of his research program. Most teachers and textbooks recognize their importance and still ignore that some of these postulates were speculation for which Bohr had no warrant or experimental evidence (for further discussion, see Chapter 12).

Consequently, in the Lakatosian framework, negative heuristic of a research program is resistant to refutation and may even be based on contradictory and inconsistent foundations. Furthermore, Lakatos (1970) shows how Bohr used the methodology of idealization (i.e., simplifying assumptions) and developed the *positive heuristic* of Bohr's program by progressing from simple to complex models, that is, from a fixed proton-nucleus in a circular orbit, to elliptical orbits, to removal of restrictions (fixed nucleus and fixed plane), to inclusion of spin of the electron (this was not in discussion in 1913), and so on until the program could ultimately be extended to complicated atoms. This illustrates quite cogently the research methodology of idealization utilized for studying physical laws in particular and complex problems in general.

The study designed by Chang and Chiu (2008) to foster argumentation is a good illustration of how the Lakatosian methodology (as contrast to other philosophers of science) can be applied in the classroom. These authors asked 70 undergraduate science and non-science majors in Taiwan to provide written arguments about four socio-scientific issues. Results showed that: (a) science majors' informal arguments were significantly better than those of non-science majors; (b) science majors made significantly greater use of analogies, while non-science majors made significantly greater use of authority; (c) both groups had a harder time changing their arguments after participating in a group discussion. According to the authors, in the study of argumentation in science education, scholars have often used Toulmin's (1958) framework of data, warrant, backing, qualifiers, claims and rebuttals. In contrast, however, in their work, the authors found that Lakatos' framework is also a viable perspective, especially when warrant and backing are difficult to discern and when students' arguments are resistant to change. This framework highlights how the "hard-core" of students' arguments about socio-scientific issues does indeed seem to be protected by a "protective belt" and thus difficult to alter.

At this stage it is important to refer to whether history of science should be rated X for science education (Brush, 1974). Following Kuhn, some scholars and even science educators have argued that detailed presentations based on history of science can present an erroneous view of science that may seem to question the certainty of scientific laws and theories. Brush (2000), a former student of Kuhn, has been considered by some circles to be opposed to the introduction of history and philosophy of science in science education. On the contrary, Brush (1978) has supported the inclusion of history of science in categorical terms:

> Of course, as soon as you start to look at how chemical theories developed and how they were related to experiments, you discover that the conventional wisdom about the empirical nature of chemistry is wrong. The history of chemistry cannot be used to indoctrinate students in Baconian methods. (p. 290)

More recently, the claim that history of science corrupts the science student and thus should not be included in the curriculum has been considered by some scholars to be "superficially bizarre" (Gooday, Lynch, Wilson & Barsky, 2008).

Chapter 2

The Role of Presuppositions, Contradictions, Controversies and Speculations versus Kuhn's "Normal Science"*

Kuhn's "Normal Science" and Science Textbooks

Kuhn (1970) is generally supposed to have been a harbinger of radical changes and even perhaps an iconoclast in the social sciences. In the case of education and especially science education, however, his influence has been more in favor of traditional approaches to teaching, and the following is an example:

> Many science curricula do not ask even graduate students to read in works not written specially for students. The few that do assign supplementary reading in research papers and monographs restrict such assignments to the most advanced courses and to materials that take up more or less where the available texts leave off. Until the very last stages in the education of a scientist, textbooks are systematically substituted for the creative scientific literature that made them possible. Given the confidence in their paradigms, which makes this educational technique possible, few scientists would wish to change it. Why, after all, should the student of physics, for example, read the works of Newton, Faraday, Einstein, or Schrödinger, when everything he needs to know about these works is recapitulated in a far briefer, more precise, and more systematic form in a number of up-to-date textbooks? *Without wishing to defend the excessive lengths to which this type of education has occasionally been carried, one cannot help but notice that in general it has been immensely effective.*
>
> (Kuhn, 1970, p. 165, emphasis added)

It could be argued that although Kuhn found the traditional textbooks to be "immensely effective", he did recognize the importance of readings that go beyond in order to familiarize the students with the "creative scientific literature". Despite some ambiguity, Kuhn (1970) did recognize that traditional science teaching "is a narrow and rigid education, probably more so than any other except perhaps in orthodox theology" (p. 166). Nevertheless, Kuhn's

* Reproduced with permission from: Niaz, M. (2010). Science curriculum and teacher education: The role of presuppositions, contradictions, controversies and speculations vs Kuhn's "normal science". *Teaching and Teacher Education, 26*, pp. 891–899.

major concern was "normal science" and he found textbooks to be good "pedagogical vehicles" to this end: "Textbooks, however, being pedagogical vehicles for the perpetuation of normal science, have to be rewritten in whole or in part whenever the language, problem-structure, or standards of normal science change" (Kuhn, 1970, p. 137). Actually, Kuhn (1963) not only recognized the importance of textbooks but went beyond by pointing out that textbooks may differ in pedagogical detail but not in substance or conceptual structure (pp. 350–351). Kuhn's views on textbooks were influenced by the seminal work of Fleck (1979/1935), and he cannot be held responsible for promoting normal science in textbooks.

Like other academic disciplines, science education has been overly influenced by Kuhn's approach to history and philosophy of science (Loving & Cobern, 2000; Matthews, 2004; Niaz, 1997; Van Berkel et al., 2000). Despite such popularity, Kuhn's ideas have also been the subject of critical scrutiny. One critic has summarized Kuhn's ideas with respect to science education, "normal science" and science textbooks in succinct terms:

> Put concisely Kuhn's view is that science education does, and should, distort the history of science. Kuhn views the goal of science education as the inculcation, in the student, of the dominant paradigm of the day. He argues that the science educator, in order to effectively inculcate that paradigm, should systematically distort the history of science ... Kuhn takes for granted that science texts do, in fact, distort the history of science, and explains this "fact" in light of their normal scientific function. According to Kuhn textbooks are designed to perpetuate normal science, which means that they are written in the language, and in accord with the principles, of the dominant paradigm of the day.
>
> (Siegel, 1979, p. 111)

It is important to note that Siegel is an important philosopher of education with a strong and sustained interest over the years in science education. However, this author does not support the thesis that either Kuhn or his "normal science" is responsible for our present-day ahistorical science textbooks and education.

Siegel (1978) has also compared Kuhn and Schwab's approaches to writing science textbooks, and recommended Schwab's application of history and philosophy of science if we want science textbooks not be "regarded as tools for inculcating in science students the principles and methods of the paradigm of the day. Rather, textbooks are to function as challengers to students" (p. 309). According to Schwab (1974), scientific inquiry tends to look for patterns of change and relationships, which constitute the heuristic (explanatory) principles of our knowledge:

> A fresh line of scientific research has its origins not in objective facts alone, but in a conception, a deliberate construction of the mind ... this conception [heuristic principle] tells us what facts to look for in the research. It tells us what meaning to assign these facts.
>
> (p. 164)

Monk and Osborne (1997) have pointed out how many science curricula have forgotten Schwab's important epistemological distinction between the methodological (experimental data) and interpretative (heuristic) components. Similarly, Stinner (1992) has argued cogently for the

> need to clarify relationships between *experiment*, *hypothesis* and *theory*, in scientific inquiry. The use of the model as a *heuristic* device would then allow an eclectic discussion of philosophical issues that would be independent of a school of thought. Moreover, repeated excursions into historical background will surely generate interest for the teacher and the student alike.
>
> (p. 14, original italics)

Similarly, the *National Science Education Standards* (NRC, 1996), in the USA, has recommended the need to understand scientific inquiry which means that students will need to have knowledge of how scientists conduct their work and understand the concepts related to the nature of science.

The importance of science-in-the-making and its relevance for students has been recognized by Osborne (2007):

> What is it about the manner in which scientific knowledge is produced that makes it reliable knowledge? How do we know what we know and why it should be valued? Understanding the epistemic aspect of science is an essential part of any comprehensive science education.
>
> (p. 179)

There is also fair amount of consensus in the science education research community for the need to write textbooks within a historical perspective in order to facilitate students' understanding of how and why experiments are performed (Chiappetta, Sethna & Fillman, 1991). Similarly, the importance of textbooks and their evaluation has also been recognized by Project 2061's *Benchmarks for Science Literacy* (AAAS, 1993) and the *National Standards for Science Education* (NRC, 1996). These reform efforts have been conducted by scientists and teachers, under the auspices of the American Association for the Advancement of Science. These documents call for not only the inclusion in textbooks of historical perspectives but also the introduction of terms meaningfully, appropriate representations of key ideas and the skilful use of models (Kesidou & Roseman, 2002).

At this stage it would be pertinent to ask: do all philosophers of science agree with Kuhn's interpretation of scientific progress? For example, according to Lakatos (1970):

> "normal science" is nothing but a research programme that has achieved monopoly. But as a matter of fact, research programmes have achieved complete monopoly only rarely and then only for relatively short periods ... The history of science has been and should be a history of competing research

> programmes (or, if you wish, "paradigms"), but it has not been and must not become a succession of periods of normal science.
>
> (p. 155)

Research Questions

This study draws attention to the role played by Kuhn's normal science, Collins' trilemma, students' understanding of nature of science (NOS) and textbook presentations, in science education and curricula, in order to provide possible answers to the following research questions:

1. What is the significance of Kuhn's "normal science" for science education and science curriculum?
2. What are the similarities between students' understanding of NOS and textbook evaluations of different topics based on historical reconstructions and NOS?
3. What is the relationship between Collins' trilemma based on Kuhn's "normal science" and the introduction of NOS in the classroom and curriculum?

This study has primarily focused on upper secondary and introductory university freshman students. Research in science education has revealed that students' epistemological views of nature of science in various parts of the world are quite similar. Similarly, textbooks written in many parts of the world have very similar empiricist frameworks. Most of the textbooks reviewed were published in the USA. It is important to note that these textbooks are used in other English-speaking countries (e.g., Canada) and as translations in languages such as Greek, Italian, Portuguese, Spanish and Turkish.

Collins' Three Horns of a "Trilemma"

Project *Beyond 2000* (Millar & Osborne, 1998), funded by the Nuffield Foundation (UK) was an attempt by science educators to elaborate a suitable model for a science curriculum, which recommended the following: students should be asked to demonstrate the capability to assess the reliability and validity of evidence. This seems to be good advice and most science educators may agree with it. Collins (2000) has commented on this in terms that are extremely important for science education:

> Stated thus, this is hopelessly ambitious, simply because the best-trained professional scientists cannot accomplish it in disputed fields. Dispute in science is, perhaps, underemphasised throughout the documents. Science education in the classroom continually misleads our future citizens by making it seem as though an hour's work at the bench can accomplish a level of certainty that took half-a-century to achieve in real life.
>
> (p. 170)

History of science shows that interpretation of experimental data (reliability and validity of evidence) is extremely difficult and frequently leads to controversies and rivalries among contending groups of scientists. Discussions based on such aspects of nature of science are important for both the future scientist and the citizenship in general. A recent appraisal of scientific controversies has concluded:

> What is not so obvious and deserves attention is a sort of paradoxical dissociation between science as actually practiced and science as perceived or depicted by both scientists and philosophers. While nobody would deny that science in the making has been replete with controversies, the same people often depict its essence or end product as free from disputes, as the uncontroversial rational human endeavour par excellence.
>
> (Machamer, Pera & Baltas, 2000, p. 3)

Collins (2000) goes on to point out the difficulties associated with the introduction of nature of science in the science curriculum, as three horns of a "trilemma": (a) excitement and thrill that comes from the ability to discover and create new knowledge and the consequent liberation from received wisdom; (b) teaching and dissemination of scientific knowledge is generally based on a dogmatic and authoritarian education, in which students must accept uncritically what they are told; (c) the necessity to teach NOS in order to appreciate and understand the different aspects of scientific development, that is, science-in-the-making (research question 3). These three horns create considerable problems for science education, and can be subsumed in the following: discovery in science can be exciting and for that we have to teach the dynamics of science-in-the-making, which is made difficult due to our present day dogmatic and authoritarian teaching of science (secondary and freshman). Collins (2000) attributes these contradictions in teaching science to the "falsified history" found in most textbooks. Furthermore, according to Collins and Pinch (1998): "Textbook history, official history, or reviewers' history are, once more, fine for scientists but damaging to those who need to understand, not the contents of science, but the way scientific facts are established" (p. 167). Research in science education in most parts of the world has recognized the importance of going beyond memorization of scientific facts and instead emphasize conceptual understanding (Linn et al., 1991). Niaz (2008a) has argued that in order to facilitate an understanding of science-in-the-making, we need to write science textbooks within a history and philosophy of science perspective. Interestingly, Osborne et al. (2003) have attributed the dogmatic and authoritarian orientation of science education to Kuhn's (1970) "normal science". It is important to note that Kuhn did question the "falsified history" in the science curriculum and textbooks, but he also found it beneficial for communicating to students the methods and achievements of science.

Students' Understanding of Nature of Science

Considerable amount of research has been conducted to show that students' understanding of how scientists work is far from satisfactory (Blanco & Niaz,

1997, 1998; Kang, Scharmann & Noh, 2005; Lederman et al., 2002; Osborne and Collins, 2001; Osborne et al., 2003; Sensevy et al., 2008; Solomon, Scott & Duveen, 1996). A detailed review of research on the subject is beyond the scope of this book. Abd-El-Khalick (2004), in a study based on 153 undergraduate and graduate students at a West Coast university in the USA, found that the majority of the students held naive views or inaccurate understanding of the following aspects of NOS: (a) the tentative, empirical, inferential, theory-laden, imaginative and creative nature of scientific knowledge; (b) social and cultural factors in theory change; (c) role of theory and prior expectations in designing and conducting experiments; (d) hierarchical view of the relationship between theories and laws; and (e) science is characterized by the use of "The Scientific Method". According to the author these results are consistent with those reported in a plethora of studies.

It is not farfetched to suggest that such an understanding is at least partially a consequence of the science curriculum and textbooks that emphasize "normal science" and ignore conflicts, controversies and presuppositions involved in interpreting experimental data. Furthermore, this raises an important issue: do we want our students to understand science as a product of the work of geniuses who know beforehand what they are going to discover? In the next section, I report findings of studies that have analyzed various topics of physical science textbooks, based on criteria related to historical reconstructions and NOS. Comparing students' understanding of NOS and textbook evaluations based on NOS criteria will facilitate a response to research question 2.

Role of Experimental Data, Presuppositions and Controversy in Physical Science Textbooks

The role of experimental data (empirical evidence) is generally considered to be crucial for accepting or rejecting a theory in science. History of science, however, shows that scientific progress has frequently involved a confrontation between the quantitative imperative and the imperative of presuppositions (Niaz, 2005a). The quantitative imperative is the view that studying something scientifically means measuring it, which was popularized by the 19th-century British physicist William Thomson (Baron Kelvin of Largs) in the following terms: "when you cannot measure it, when you cannot express it in numbers, your knowledge is of a meagre and unsatisfactory kind" (Thomson, 1891, pp. 80–81). This methodological rule has been referred to as the "Kelvin dictum" by Merton, Sills and Stigler (1984) and is still widely accepted in both science and science education. Despite the popularity of the Kelvin dictum, it is important to recognize that progress in science, starting at least from Galileo (1564–1642), has gone through a constant process of conflict and controversy, in which the quantitative imperative has been confronted by the imperative of presuppositions. In other words, not all scientific theories and laws have been strictly based on quantitative data (despite the rhetoric of the quantitative imperative), but rather there has been a continual critical appraisal of data based on hypotheses and presuppositions (Machamer et al., 2000). Based on the existing knowledge in a field of research, a scientist formulates the guiding assumptions (Laudan, Laudan & Donovan, 1988), presuppositions (Holton, 1978a, 1978b) and "hard-core" (Lakatos, 1970) of the research program that constitutes the imperative of presuppositions, which is not abandoned in the face of anomalous data.

In this section I briefly summarize some historical episodes in the history of science, to demonstrate that experimental data (quantitative imperative) by itself does not constitute progress in science. On the contrary, it is the critical appraisal of the data by the scientific community (peer review) that facilitates understanding. Scientists generally ignore the dynamics of scientific change, namely, the difficulties and controversies involved in understanding data, and this inevitably leads to the formulation of Kuhn's "normal science". The next step is almost a logical consequence in which the science curricula and textbooks adopt the same epistemological stance.

Thomson's Experiments: Cathode Rays as Charged Particles or Waves in the Ether

Thomson's experiments were conducted against the backdrop of a conflicting framework. Thomson (1897) explicitly points out that his experiments were conducted to clarify the controversy with regard to the nature of cathode rays, that is, charged particles or waves in the ether. Resolution of this controversy constituted a heuristic principle (cf. Schwab, 1974) for Thomson's research program and its inclusion in the textbooks could help students to understand why scientists do experiments. Niaz (1998) has shown that of the 23 general chemistry textbooks analyzed (published in the USA) only two made a simple mention of this controversy. Rodríguez and Niaz (2004a) have shown that of the 41 general physics textbooks analyzed only two simply mentioned the controversy and two made a satisfactory presentation.

Thomson–Rutherford Controversy: Single/Compound Scattering of Alpha Particles

In the very first paragraph of his famous article in the *Philosophical Magazine*, Rutherford (1911) starts on controversial note: "It has generally been supposed that the scattering of a pencil of alpha or beta rays in passing through a thin plate of matter is the result of a multitude of small scatterings by the atoms of matter traversed" (p. 669). This, of course, referred to the experimental work of Crowther (1910), a colleague of Thomson. Rutherford had the experimental data to postulate his model of the nuclear atom as early as June 1909, and yet he did not do so until March 1911. What happened between June 1909 and March 1911 is important not only for historians and philosophers of science, but also for science teachers. Soon after Geiger and Marsden (1909) published their results, Thomson and colleagues also started working on the scattering of alpha particles in their laboratory. Although experimental data from both laboratories were similar, interpretations of Thomson and Rutherford were entirely different. Thomson propounded the hypothesis of *compound scattering*, according to which a large-angle deflection of an alpha particle resulted from successive collisions between the alpha particles and the positive charges distributed throughout the atom. Rutherford, in contrast, propounded the hypothesis of *single scattering*, according to which a large-angle deflection resulted from a single collision between the alpha particle and the massive positive charge in the nucleus. The rivalry led to a bitter dispute between the proponents of the two hypotheses. Rutherford even charged that Crowther (1910), a colleague of Thomson, had "fudged" the data in order to provide support for Thomson's model of the atom (Wilson, 1983). Rutherford's dilemma: on the one hand he was entirely convinced and optimistic that his model of the atom explained experimental findings better, and on the other hand it seems that the prestige, authority and even perhaps some reverence for his teacher made him waver. However, in a letter to Schuster (Secretary of the Royal Society), written about 3 years later (February 2, 1914), Rutherford is much more forceful:

> I have promulgated views on which J.J. [Thomson] is, or pretends to be, sceptical. At the same time I think that if he had not put forward a theoretical atom himself, he would have come round long ago, for the evidence is very strongly against him. If he has a proper scientific spirit I do not see why he should hold aloof and the mere fact that he was in opposition would liven up the meeting.
>
> (Reproduced in Wilson, 1983, p. 338)

This makes interesting reading. A science student may wonder why Thomson and Rutherford did not meet over dinner (they were well known to each other) and decide in favor of one or the other model. Progress in science is, however, much more complex. Both Thomson and Rutherford stuck to their presuppositions. Again, a student may wonder how Rutherford could doubt the "proper scientific spirit" of none else but the world master in the design of atomic models. These issues, if discussed in class and textbooks, could make science much more human and attractive. Niaz (1998) has reported that of the 23 general chemistry textbooks analyzed none described satisfactorily or mentioned this controversy. Rodríguez and Niaz (2004a) have reported that of the 41 general physics textbooks analyzed, two described satisfactorily and two made a simple mention. All textbooks analyzed were published in the USA.

Bohr's Model of the Atom: Based on Inconsistent Foundations and a Deep Philosophical Chasm

An important aspect of Bohr's model of the atom is the presence of a deep philosophical chasm: that is, in the stationary states, the atom obeys classical laws of Newtonian mechanics; on the other hand, when the atom emits radiation, it exhibits discontinuous (quantum) behavior. Based on these and other arguments, Bohr's 1913 article, in general, had a fairly adverse reception in the scientific community. Rutherford, although no philosopher of science, was the first to point this out, when he wrote to Bohr on March 20, 1913:

> the mixture of Planck's ideas with the old mechanics makes it very difficult to form a physical idea of what is the basis of it all ... How does the electron decide what frequency it is going to vibrate at when it passes from one stationary state to another?
>
> (Reproduced in Holton, 1993, p. 80)

Lakatos (1970) has argued that Bohr employed a methodology used frequently by scientists in the past and perfectly valid for the advancement of science:

> *some of the most important research programmes in the history of science were grafted on to older programmes with which they were blatantly inconsistent.* For instance, Copernican astronomy was "grafted" on to Aristotelian physics, Bohr's programme on to Maxwell's. Such "grafts" are irrational for the justificationist and for the naive falsificationist, neither of whom can countenance growth on inconsistent foundations ... As the young grafted programme strengthens, the peaceful co-existence comes to an end, the symbiosis becomes competitive and the champions of the new programme try to replace the old programme altogether.
>
> (p. 142, original italics)

Niaz (1998) has reported that of the 23 general chemistry textbooks analyzed, four simply mentioned this heuristic principle and two made a satisfactory presentation. Rodríguez and Niaz (2004a) have reported that of the 41 general physics textbooks analyzed, only one described satisfactorily that Bohr's incorporation of Planck's "quantum of action" was based on an inconsistent foundation and represented a "deep philosophical chasm".

Millikan–Ehrenhaft Controversy: Determination of the Elementary Electrical Charge

A historical reconstruction of the oil drop experiment that led to the determination of the elementary electrical charge (e) shows both the controversial nature of the experiment then (1910–1925) and that the experiment is difficult to perform even today (Jones, 1995). After almost 90 years, historians and philosophers of science do not seem to agree on what really happened (Niaz, 2005b). Despite these difficulties, most chemistry and physics textbooks consider the oil drop experiment to be a simple, classic and beautiful experiment, in which Robert A. Millikan (1868–1953), by an exact experimental technique, determined the elementary electrical charge.

An important aspect of Millikan's experiments (University of Chicago) is that he clearly formulated the guiding assumptions (hard-core, cf. Lakatos, 1970) of his research program from the very beginning. Felix Ehrenhaft (1879–1952), Millikan's major critic, did most of his experimental work at the University of Vienna and was considered in 1910 (when the controversy started) a fairly well established figure in the European scientific community. In Ehrenhaft's (1910) first major criticism, he closely scrutinized Millikan's (1910) data. He recalculated the charge on each drop from each of Millikan's observations separately. Millikan (1910), in contrast, used average values of times of ascent and descent, measured on different droplets. Ehrenhaft's calculations produced a large spread of values of the elementary electrical charge, ranging from 8.60×10^{-10} esu to 29.82×10^{-10} esu. Furthermore, Ehrenhaft showed how Millikan's method led to paradoxical situations. Holton (1978a) provides the following insight on the impasse: "It appeared that the same observational record could be used to demonstrate the *plausibility of two diametrically opposite theories*, held with great conviction by two well equipped proponents and their respective collaborators" (pp. 199–200, emphasis added). Ehrenhaft wrote about a dozen articles in the following 4 years, all implicitly aimed at discrediting Millikan's measurements. Millikan also wrote extensively and rebutted Ehrenhaft's criticisms. Millikan's data showed that all drops had a charge that was an integral multiple of the elementary electrical charge, thus providing evidence for the existence of the electron. Research literature at the time considered Rutherford and Geiger's (1908) value of $e = 4.657 \times 10^{-10}$ esu as the most probable value. Millikan's data provided a value of *e* very close to this. Ehrenhaft, too, obtained data that he interpreted as integral multiples of the elementary electrical charge (*e*). Nevertheless, his argument was precisely that there were many drops that did not lead to an integral multiple of *e*.

A new dimension to the controversy was added by Holton's (1978) discovery of Millikan's laboratory notebooks in his archives at the California Institute of Technology, Pasadena. These notebooks have raw data on 140 drops and some of the data reduction procedures used in his *Physical Review* article (Millikan, 1913). The Millikan notebooks (October 28, 1911 to April 16, 1912—about 175 pages) are indeed a rare opportunity to see a scientist working in his laboratory.

Now let us turn to the actual publication, in which Millikan (1913) meticulously presented complete data on 58 drops and emphasized that all of the drops experimented upon had been included. The laboratory notebooks tell us that there were 140 drops and the published results are emphatic that there were 58 drops. What happened to the other 82 drops? Herein lies the crux of the difference between the methodologies of Ehrenhaft and Millikan. What was the warrant under which Millikan discarded more than half of his observations? The answer is simple but not found frequently in the research literature, and much less in science textbooks. Millikan's guiding assumptions provided the warrant.

Barnes, Bloor and Henry (1996) have raised another important issue with respect to the oil drop experiment: "we should avoid the inference to the rightness of Millikan's theory from the fact that it works" (p. 30). Based on Millikan's notebooks and data reduction procedures it is quite clear that his theory was not right even for some of his own data. Niaz (2005b) has argued in the context of this experiment that scientific theories need not be evaluated on the basis of rightness or wrongness, but instead on their heuristic/explanatory power. Furthermore, according to Lakatos (1970), scientific theories are tentative. For example, was Thomson's (1897) theory right then and became wrong after Rutherford (1911) published his model of the atom? Rutherford's model increased the heuristic power of the theory, just as Millikan's determination of the elementary electrical charge increased the heuristic power of the atomic theory even further (for details, see Niaz, 2009a). On the whole, Barnes et al. (1996) have facilitated our understanding of the experiment and thus played a role envisaged by Fuller (2000): "sociologists can step into the breach when philosophers cannot decide among themselves which methodology best explains a certain historical episode of scientific theory choice" (p. 141).

Niaz (2000a) has reported that of the 31 general chemistry textbooks analyzed none mentioned the Millikan–Ehrenhaft controversy. Similarly, Rodríguez and Niaz (2004b) have reported that of the 43 general physics textbooks analyzed none mentioned the controversy. Some textbooks explicitly denied that the drops studied by Millikan had fractional charges, i.e., a charge unequal to an integer times the electron charge.

Kinetic Theory of Gases: Inconsistent Nature of Maxwell's Research Program

J.C. Maxwell's research program with respect to kinetic theory is yet another example of a program progressing on inconsistent foundations. Among other assumptions, Maxwell's (1860) paper was based on "strict mechanical principles" derived from Newtonian mechanics and yet at least two of Maxwell's simplifying assumptions (referring to movement of particles and consequent generation of pressure) were in contradiction with Newton's hypothesis explaining the gas laws based on repulsive forces between particles. Newton provided one of the first explanations of Boyle's law in his *Principia* (1687) in the following terms: "If a gas is composed of particles that exert repulsive forces on their neighbors, the magnitude of force being inversely as the distance, then the pressure will be inversely as the volume" (Brush, 1976, p. 13). Due to Newton's vast authority, Maxwell even in his 1875 paper reiterated that Newtonian principles were applicable to unobservable parts of bodies (Achinstein, 1987, p. 418). Brush (1976) has pointed out the contradiction explicitly: "Newton's laws of mechanics were ultimately the basis of the kinetic theory of gases, though this theory had to compete with the repulsive theory attributed to Newton" (p. 14).

Niaz (2000b) has reported that of the 22 general chemistry textbooks analyzed none referred to the inconsistent nature of Maxwell's research program. Many textbooks explicitly invoke Newtonian mechanics along with Maxwell's presentation of the kinetic theory, without realizing an inherent contradiction. Similarly, Rodríguez and Niaz (2004c) have reported that of the 30 general physics textbooks analyzed none referred to the contradiction. All textbooks analyzed were published in the USA.

Development of the Periodic Table: Mendeleev's Contribution—Theory or an Empirical Law?

There seems to be considerable controversy among philosophers of science with respect to the nature of Mendeleev's contribution, namely, what exactly was he trying to do with all the data available. Mendeleev's (1879, 1889) own ambivalence notwithstanding, the historical reconstruction shows that Mendeleev's ingenuity consisted precisely of not only recognizing that the periodic table was a "legitimate induction from the verified facts" but that there was a reason/cause/explanation for this periodicity, namely, the atomic theory (Niaz, Rodríguez & Brito, 2004). In other words, scientists do not decide beforehand that their contribution would be empirical/theoretical, but rather the scientific endeavor inevitably leads them to "speculate" with respect to underlying patterns of what they observe. Mendeleev's case is an eloquent example of this dilemma. Brito et al. (2005) have reported that of the 57 general chemistry textbooks (published in the USA) analyzed, none made a satisfactory presentation and 52 simply ignored the issue.

Brief Review of Evaluation of Textbooks Based on a Historical Framework

Leite (2002) analyzed five high school physics textbooks published in Portugal on criteria such as historical experiments, analyses of data from historical experiments, integration of historical references within the text, use of original historical sources, evolution of science and socio-political context in scientific research. The author concluded that the historical content included in the textbooks hardly provided students with an adequate image of science and the work of the scientists.

Justi and Gilbert (1999) analyzed high school chemistry textbooks (nine from Brazil and three from the UK) to study the presentation of atomic models. These authors report the use of hybrid models based on various historical developments, such as Ancient Greek, Dalton, Thomson, Rutherford, Bohr and quantum mechanics (Schrödinger's equation). The authors concluded: "Hybrid models, by their very nature as composites drawn from several distinct historical models, do not allow the history and philosophy of science to make a full contribution to science education" (p. 993).

A historical reconstruction shows that it was the acceptance of the atomic-molecular theory that facilitated an understanding of "amount of substance" and its unit the "mole" (Padilla & Furio-Mas, 2008). In a study based on 30 general chemistry textbooks (published in the USA), these authors found that a majority of the textbooks present an ahistoric and aproblematic interpretation of this topic.

Abd-El-Khalick, Waters and Le (2008) have drawn attention to the importance of including NOS in high school chemistry textbooks. These authors analyzed 14 textbooks including five "series" spanning one to four decades, with respect to the following NOS aspects: empirical, tentative, inferential, creative, theory-driven, myth of the scientific method, nature of scientific theories and laws and the social and cultural embeddedness of science. Results from this study revealed that chemistry textbooks fared poorly in their representation of NOS, which led the authors to conclude, "These trends are incommensurate with the discourse in national and international science education reform documents" (p. 1).

Based on content analysis of school chemistry textbooks and syllabi, Van Berkel et al. (2000) have identified the dominant school chemistry as a form of normal science education (NSE), which is in turn based on Kuhn's "normal science". These authors have alluded to the dangers of NSE as it is isolated from history and philosophy of science, is narrow and rigid and instills a dogmatic attitude.

All studies (related to textbooks) reviewed in this section had the following characteristics: (a) most of them were published in major science education journals indexed in *Social Science Citation Index* (Thomson-Reuters); (b) all of them have a protocol or framework for the analyses; (c) most of them provide procedures and estimates of reliability.

At this stage it may be argued that research in science in the past may have gone through rivalries and contradictions among scientists. However, present-day scientists generally follow the scientific method. Recently, Jenkins (2007) has traced the origin of the scientific method as a political construct in the 19th century, and that it is at odds with developments in history and philosophy of science in the 20th century. Furthermore, researchers in science education have explicitly exposed the myth of the scientific method (Lederman et al., 2002; Millar, 1989; Millar & Driver, 1987; Niaz, 1994). Despite such advice, textbooks continue to emphasize the importance of the scientific method both in science and science education (Russo & Silver, 2002).

Now let us see what an experimental physicist working on cutting-edge research has to say about research methodology in present-day science. Martin Perl, Nobel Laureate in physics (1995), in his search for the experimental isolation of the fundamental particle (quark) elaborated a philosophy of speculative experiments:

> Choices in the design of speculative experiments [cutting-edge] usually cannot be made simply on the basis of pure reason. The experimenter has to base her or his decision partly on what feels right, partly on what technology they like, and partly on what aspects of speculation they like.
>
> (Perl & Lee, 1997, p. 699)

A critical reader may wonder, what does all this have to do with Kuhn's "normal science"? To recapitulate, Kuhn (1970) is explicit in requiring not only that textbooks be rewritten according to the prevailing standards of normal science but that "there are even good reasons why, in these matters, they should be systematically misleading" (p. 137). A review of studies that deal with analyses of textbooks presented in this section clearly shows that textbooks are "systematically misleading" our future generations of citizens and scientists, by providing an ahistoric account of progress in science in which presuppositions, contradictions, conflicts and controversies have no role to play. If we follow Kuhn's advice, all reform efforts at improving the science curriculum (teacher training and including textbooks) are misdirected and needless, as dissemination of normal science will facilitate more interest and understanding of science (research question 1). The crux of the issue at stake can be understood better in the context of

Collins' (2000) trilemma: our dogmatic and authoritarian science education (normal science) can be counteracted to make science more attractive and appealing to students by including different aspects of science-in-the-making, namely historical, philosophical and sociological context of scientific progress (research question 3).

Resolution of Collins' Trilemma and Kuhn's Normal Science

Hodson (1985), a prominent science education researcher, has explicitly endorsed Kuhn's "normal science":

> The pupil's first priority is to *learn* currently accepted theories and to apply them to appropriate phenomena and in appropriate situations ... Much of the laboratory work in school should concentrate on theory illustration and investigation, rather in the manner of Kuhn's normal science.
>
> (p. 43)

Let us pause for a moment to consider the Kuhnian and the Lakatosian approaches to teaching, for example, the structure of the atom. These approaches were not suggested by Kuhn or Lakatos themselves, but would rather follow from their endorsement of "normal science" or "science in the making". Most high school and general chemistry textbooks refer to the work on the subject by scientists such as Thomson, Rutherford and Bohr. From a Kuhnian point of view, what is important in Thomson's experiments is that they have contributed to the development of the currently accepted theory in the field, namely, that the charge-to-mass ratio for cathode rays is a constant independent of the gas used in the discharge tube. On the other hand, from a Lakatosian point of view, it is essential to point out that there were "competing frameworks of understanding that clash in the face of evidence", namely, a determination of the charge-to-mass ratio for the cathode ray particles would help identify it either as an ion or some other charged particle. The experimental work of Rutherford is important from the Kuhnian point of view as it postulated the existence of a positively charged nucleus and surrounding electrons. On the contrary, a Lakatosian point of view would emphasize the fact that Rutherford's experiment provided evidence against a competing framework, namely, Thomson's model of the atom. Once again, from the Kuhnian point of view, perhaps it is important that Bohr developed a model of the hydrogen atom which allowed him to explain the fact that the frequencies emitted were in agreement with the experimental work of Balmer's and Paschen's series. On the contrary, Lakatos has emphasized that Bohr's major concern was to explain the paradoxical stability of the Rutherford atom, that is, a competing framework that was not entirely consistent. For Bohr's research program:

> The background problem was the riddle of how Rutherford atoms (that is, minute planetary systems with electrons orbiting round a positive nucleus)

> can remain stable; for, according to the well-corroborated Maxwell-Lorentz theory of electromagnetism they should collapse. But Rutherford's theory was well corroborated too.
>
> (Lakatos, 1970, p. 141)

The emerging and the inconsistent nature of Bohr's research program is captured by Lakatos (1970) in truly picturesque terms: "for Bohr's atom sat like a baroque tower upon the Gothic base of classical electrodynamics" (p. 142). Interestingly, Lakatos (1971) goes so far as to suggest that from a Popperian point of view, Bohr's 1913 paper should never have been published, as it was "inconsistently grafted on to Maxwell's theory" (p. 113). It appears then that emphasizing a Kuhnian perspective of "normal science" (cf. Hodson, 1988a), in which the students, for example, "*learn*" variously about the charge-to-mass ratio of the cathode ray particles, existence of a positively charged nucleus and the surrounding electrons, a model of the hydrogen atom which explained the experimentally observed frequencies, and so on, would deprive them of the dynamics of "science in the making". Niaz et al. (2002) have reported a study in which these ideas were used to facilitate freshman students' conceptual understanding of atomic structure. On the contrary, many teachers and most textbooks follow an approach that approximates to Kuhn's "normal science". However, it appears that the Kuhnian approach leaves out what really happens, that is the "how" and "why" of scientific progress. On the other hand, the Lakatosian perspective would enable students to understand that scientific progress is subsumed by a process involving conflicting frameworks, based on processes that require the elaboration of rival hypotheses and their evaluation in the light of new evidence (research question 3; for further discussion of these issues, see Niaz, 2008c). Similarly, Popper (1970) voiced a concern with respect to teaching "normal science" both at the secondary and university level.

Conclusion

Given Kuhn's popularity in science education for the last 25 years, it is important to subject some of his propositions to a critical appraisal with respect to "normal science" in the context of reform efforts (*Project 2061*, *Beyond 2000*), suggesting that science education be brought more in consonance with how science is practiced by the scientists. Kuhn considered textbooks to be good "pedagogical vehicles" for the perpetuation of "normal science". Collins (2000) has pointed out a fundamental contradiction (research question 1) with respect to what science could achieve (discover and create new knowledge) and how we teach science (dogmatic and authoritarian). Despite the reform efforts, students (secondary and university) still have naive views about the nature of science in which experimental data unambiguously lead to the formulation of laws and theories. Review of the literature based on textbook analyses shows almost a complete lack of understanding of the role played by presuppositions, contradictions, controversies and speculations in scientific progress. Furthermore, students' understanding of NOS is quite similar to that of the textbook (research question 2). Kuhn's advice based on "normal science" would seem to suggest that the science curriculum need not appeal to the imagination and creativity of the students.

A possible solution to Collins' trilemma (research question 3) is provided by the comparison of teaching approaches based on Kuhnian and Lakatosian perspectives of history and philosophy of science. It appears that the Kuhnian approach leaves out what really happens, that is the "how" and "why" of scientific progress. On the other hand, the Lakatosian perspective would enable students to understand that scientific progress is subsumed by a process that involves conflicting frameworks (dispute in science, according to Collins, 2000), based on processes that require the elaboration of rival hypotheses and their evaluation in the light of new evidence. It is plausible to suggest that the teacher, by "unfolding" the different episodes (based on historical reconstructions), can emphasize and illustrate how science actually works (tentative, controversial, rivalries, alternative interpretations of the same data), and this will show to the students that they need to go beyond "normal science" as presented in their textbooks. Such teaching experiences can provide teachers an opportunity to fire the imagination of the students in order to go beyond "normal science" and provide them with different scenarios of science-in-the-making. Jenkins (2007) has expressed this concern in cogent terms:

> school science thus faces a number of challenges. It needs to reflect important philosophical, conceptual, and methodological differences between at least the basic sciences and to develop pedagogical strategies that present scientific inquiry in terms that accommodate the creative and imaginative as well as the logical.
>
> (p. 277)

A historical reconstruction of the different topics of physical science shows that although experiments are important, interpretation of the data is even more important. In order to develop their research program scientists rely on their guiding assumptions, which inevitably leads to conflicts and controversies. Even when textbooks present historical details of a scientist's work, it invariably is in the form of their pictures, laboratory instruments, year and place of work and anecdotes. Such presentations lack a framework based on history and philosophy of science and hence provide little additional insight to students as to how scientists work and theories are developed. If we want our students to understand and really scrutinize scientific practice, then a revision of the science curriculum and teacher training is necessary.

Finally, let us go back to Collins' (2000) critique of science education and the reform efforts: "Dispute in science is, perhaps, underemphasized throughout the documents" (p. 170). This is the crux of the issue, if we want to change the science curriculum. This study formulated three research questions in order to throw light on this issue. If we follow the chain of thought provided by the different sections of this study, it is plausible to construct the following sequence of arguments: Kuhn's "normal science" manifests itself in the science curriculum and textbooks through the scientific method—teaching normal science leads to memorization of science content with little understanding—eliminating normal science would facilitate the inclusion of controversies (Collins' disputes)—unfolding of the historical episodes based on rival interpretations can provide students a glimpse of what science is all about. This reconstruction facilitates guidelines for elaborating strategies for teacher training: (a) in the traditional science curriculum, "normal science" generally conveys the message that this is true and so you must accept and learn. As students are not provided any particular reason for accepting a theory, they end up memorizing it, reminiscent of Schwab's (1974) "rhetoric of conclusions". Teachers implement this in the classroom through a recipe-like scientific method, generally espoused by textbooks and accepted uncritically by teachers; (b) in the new science curriculum, students could be exposed to historical episodes by highlighting the difficulties involved in understanding experiments, which lead to controversies based on different interpretations of the same data. This will provide students with reasons and background for understanding and accepting a theory. The difference between the traditional and new science curriculum is quite clear: in the former you are told what to accept, whereas in the latter students are not only given the reasons for accepting a theory but also the teaching strategy, based on arguments and counter-arguments (Niaz et al., 2002). Interestingly, Collins and Pinch (1993, p. 151), in an attempt to put the golem to work, provide an example from a laboratory experiment in which students are asked to determine the boiling point of water. As most teachers must have found, nobody would get the value of 100 °C. According to the authors this provides an opportunity to pause and reflect as to the reasons for this and thus go beyond the traditional science classroom. In a review of curriculum materials based on *Project 2061*, Kesidou and Roseman (2002) have specifically emphasized that curriculum material needs to focus on sound instructional strategies.

Next Chapter

In the "how" and "why" of scientific progress it is important to emphasize the role of presuppositions, contradictions, controversies and speculations, which contrast with Kuhn's "normal science". The teacher, by "unfolding" the different episodes (based on historical reconstructions), can emphasize and illustrate how science actually works. A major argument in favor of mixed methods (integrative) research programs is that it provides a rationale for hypotheses, theories, guiding assumptions and presuppositions to compete and provide alternatives (subject of Chapter 3). Similar to the physical sciences, this proliferation of hypotheses in education, leads to controversies and rivalries, and thus facilitates the decision-making process of the scientific community.

Chapter 3

A Rationale for Mixed Methods (Integrative) Research Programs in Education*

Introduction

It seems that after the rhetoric of the paradigm wars (Gage, 1989; Guba, 1990; Lincoln, 1989) research in education is now heading toward a new research program with some degree of consensus. Johnson and Onwuegbuzie (2004) have advocated a trilogy of research paradigms, namely qualitative research, quantitative research and mixed methods (integrative) research and outlined the *fundamental principle of mixed research* in the following terms: "According to this principle, researchers should collect multiple data using different strategies, approaches, and methods in such a way that the resulting mixture or combination is likely to result in complementary strengths and nonoverlapping weaknesses" (p. 18). Despite some consensus there has also been controversy with respect to the "evidence based research" or the "gold standard" (randomized controlled experiments) as recommended by the National Research Council (NRC, 2002). Phillips (2005a) has steered away from the extreme views and critiqued educational research for not providing real examples. He compares the present stage of philosophy of educational research to the philosophy of science of six decades ago, when in-depth historical studies of Popper, Kuhn, Lakatos, Cartwright, Gallison, Sober and other contemporary philosophers of science started to appear. Phillips and Burbules (2000) consider educational research as a fallible enterprise that attempts to construct viable warrants or chains of arguments that draw upon diverse bodies of evidence that support the assertions being made.

According to Johnson and Onwuegbuzie (2004), emergence of the new paradigm (mixed methods research) is possible as both qualitative and quantitative (i.e., postpositivists, Phillips & Burbules, 2000) researchers have reached basic agreement on the following philosophical points: (a) what appears reasonable can vary across persons; (b) theory-ladenness of observations; (c) same experimental data can be explained by different theories; (d) the Duhem–Quine thesis; (e) empirical evidence does not provide conclusive proof; and (f) attitudes, beliefs and values of the researchers influence their findings, so that fully

* Reproduced with permission from: Niaz, M. (2008d). A rationale for mixed methods (integrative) research programmes in education. *Journal of Philosophy of Education, 42*(2), 287–305.

objective and value-free research is a myth (for a similar argument, see Smeyers, 2006, p. 479). This is an ambitious agenda and if researchers on both sides of the divide subscribe to it, then it may provide grounds for not only the emergence but rather proliferation of new paradigms (preferably research programs).

In order to justify the emergence of the new paradigm (mixed methods), Johnson and Onwuegbuzie (2004) have presented a detailed analysis of the weaknesses and strengths of qualitative, quantitative and mixed methods and finally embraced pragmatism (Peirce, James & Dewey) as a philosophical alternative. The objective of this study is to show that, based on contemporary philosophy of science, there are other alternatives. I will first present a critique of the philosophical underpinnings of the research paradigm of Guba and Lincoln (considered to be the leading qualitative researchers in the world), which is essentially based on Kuhn's (1970) philosophy of science. Next, I will present an outline of the contemporary philosophy of science based on Lakatos (1970, 1971), Giere (1999), Cartwright (1983, 1989, 1999) and Holton (1978a, 1978b, 1998), which shows that progress in science is characterized by conflicting and rival research programs (not paradigms, cf. Kuhn, 1970), integrating both quantitative and qualitative aspects. Positivist historians and philosophers of science emphasized the experimental data (quantitative) and generally ignored the interpretations (qualitative), namely, how do researchers come to understand the significance of their data? In quantitative educational research this came to be characterized by the Null Hypothesis Significance Testing Procedure (NHSTP). According to Ratnesar and Mackenzie (2006), NHSTP is still considered to be the sine qua non of scientific research in many methods courses. Contemporary philosophers of science, in contrast, provide greater insight by showing that observations are theory-laden and it is precisely the integration of the data and the presuppositions (hypotheses, guiding assumptions, hard-core of the research program) that facilitate interpretations and conceptual understanding in science. Finally, this will help to conceptualize qualitative, quantitative and mixed research programs (not paradigms) as rivals and not necessarily a displacement of one by the other, but rather integration. This insight is important, as research programs are not right or wrong, but rather are important in the degree to which they provide heuristic (explanatory) power.

A Critical Appraisal of Guba and Lincoln's Research Program

One of Lincoln's main theses is that there is a *paradigm revolution* in many disciplines (e.g., natural science, brain theory, philosophy, linguistics, politics). The concept of paradigm revolution is based on a Kuhnian interpretation, summarized in the following terms:

> As scientists proceed with their work, anomalies occasionally occur. These anomalies represent phenomena which cannot be either understood or explained by current and existing theories. For some period of time, the anomalies are simply understood as anomalies, until the sheer number and

> weight of them begins to stretch a given theory's credibility to the point of collapse ... Sooner or later, however, the anomalies begin to take on the appearance of regularities, and a crisis is precipitated, according to Kuhn. In simplest terms, the crisis revolves about whether the scientific community will reject traditional theory and begin anew to construct theory which accounts for the anomalies, or whether it will remain tied to classical theory. A battle ensues between the classicists and the emergent theoreticians.
>
> (Lincoln, 1989, pp. 60–61)

This is a fairly standard Kuhnian account of scientific progress, which helps Lincoln to suggest that in education, the qualitative research paradigm must displace the quantitative. Alexander (2006) has presented a similar critique. A reconstruction of many episodes in the history of science, however, presents a different picture. Niels Bohr (1913), in order to present his model of the atom based on the emergent paradigm (quantum theory), not only did not discard the classical electrodynamical theory of Maxwell, namely, the old paradigm (as required by Kuhn), but rather, "inconsistently grafted [his model] on to Maxwell's theory" (Lakatos, 1971, p. 113). The emerging and the inconsistent nature of Bohr's research program is captured by Lakatos (1970) in truly picturesque terms, "for Bohr's atom sat like a baroque tower upon the Gothic base of classical electrodynamics" (p. 142). Actually, Lakatos (1970) has emphatically endorsed the development of scientific theories on inconsistent foundations:

> Indeed, some of the most important research programmes in the history of science were grafted on to older programmes with which they were blatantly inconsistent. For instance, Copernican astronomy was "grafted" on to Aristotelian physics, Bohr's programme on to Maxwell's. Such "grafts' are irrational for the justificationist and for the naive falsificationist, neither of whom can countenance growth on inconsistent foundations.
>
> (p. 142)

This shows that the qualitative research paradigm does not have to necessarily displace the quantitative paradigm, but rather the two can "live" together. At this stage it is important to clarify that the analogy between physical science and research in education is valid, as an essential part of Guba and Lincoln's research program is based on such analogies. Furthermore, Thagard (1992) has referred to this as "cross-domain analogies" such as Darwin's use of Malthus's notions of population increase in humans to develop his theory of natural selection. Similarly, Gholson and Barker (1985) have drawn a parallel between the historical development of physics and psychology.

Kuhnian philosophy has been a major source of inspiration for qualitative researchers and it seems that the following aspects of his philosophy have played an important role: (a) it presupposes subjectivity as an integral part of the scientific process, once thought to be wholly objective; (b) it asserts that different paradigms are incommensurate because their core beliefs are resistant to change; (c) paradigms do not merge over time, rather they displace each other after periods of chaotic upheaval or scientific revolution. Kuhnian displacements are not subtle events. They are described as cataclysmic clashes in which losers languish and victors flourish. Of these three aspects, there is some consensus in the educational research community on the first, namely, the degree to which a researcher can be objective (cf. Johnson & Onwuegbuzie, 2004, for philosophical points on which there is consensus, as reported in the first section). The other two aspects are the subject of considerable controversy (Barker & Gholson, 1984; Friman, Allen, Kerwin & Larzelere, 1993; Malone, 1993; Niaz, 1997; Reese & Overton, 1972).

According to Barker and Gholson (1984), Kuhn's incommensurability thesis (revolutions interspersed with periods of "normal science" and that rational debate about competing paradigms is almost impossible) is interpreted by many researchers to signify that any one science can accommodate only one paradigm. This contentious aspect of Kuhn's philosophy may be the reason so many qualitative researchers have interpreted the popularity of their research program as necessarily based on the displacement of the quantitative research program. Lincoln (1989) and some other qualitative researchers have imbibed so heavily from Kuhn's incommensurability thesis that they ignore its criticism by other philosophers of science (cf. Laudan, Laudan & Donovan, 1988 for a comprehensive critique).

Before going on to the next section, it is important to note that following Galileo's method of idealization (considered to be at the heart of all modern physics by Cartwright, 1989, p. 188) scientific laws, being epistemological constructions, do not describe the behavior of actual bodies. Galileo's law of free fall, Newton's laws, gas laws, Piaget's developmental stages—they all describe the behavior of ideal bodies that are abstractions from the evidence of experience and the laws are true only when a considerable number of disturbing factors, itemized in the *ceteris paribus* clauses, are eliminated (cf. Ellis, 1991; Matthews, 1987; McMullin, 1985; Niaz, 1999a, 1999b). McMullin (1985) considers the manipulation of variables (disturbing factors) as an important characteristic of Galilean idealization:

> The move from the complexity of nature to the specially contrived order of the experiment is a form of idealization. The diversity of causes found in Nature is reduced and made manageable. The influence of impediments, i.e., causal factors which affect the process under study in ways not at present of interest, is eliminated or lessened sufficiently that it may be ignored.
>
> (p. 265)

In order to "prove" his law of free fall, Galileo should have presented empirical evidence to his contemporaries by demonstrating that bodies of different weight (but of the same material) fall at the same rate. If the leaning tower of Pisa mythical experiment (Segre, 1989) was ever conducted, it would have shown Galileo to be wrong. According to Pascual-Leone (1978), a leading cognitive psychologist, empirical computation of the value of *s* (distance) as a function of the variable *t* (time), "where vacuum and other *simplifying assumptions* are not satisfied" (emphasis added, p. 28), would lead to a rejection of Galileo's law of free fall, in very much the same way as Piaget's theory is rejected on empirical grounds. Similarly, Lewin (1935, p. 16) has questioned the excessive dependence on empirical data, referred to as "the statistical way of thinking", which underlies the conflict between the Aristotelian and Galilean modes of thought in contemporary psychology.

In the next section, examples drawn from the history of physical science are presented to demonstrate how there was a confrontation between the quantitative imperative and the latent imperative of presuppositions (qualitative interpretations), which inevitably led to controversy, and thus facilitated integration. The quantitative imperative is the view that studying something scientifically means measuring it, namely, collecting quantitative data (cf. Michell, 2003, 2005). Niaz (2005a) has argued that despite the dominance and popularity of the quantitative imperative, progress in science has frequently involved a confrontation between the quantitative imperative and the imperative of presuppositions.

Framework for Integration Based on Contemporary Philosophy of Science

Newton and the Law of Gravitation

According to Lakatos (1970) Newton's law of gravitation is one of the "best corroborated scientific theory of all time" (p. 92). Feynman (1967) endorses the view that it is "the greatest generalization achieved by the human mind" (p. 14). In spite of such impressive credentials, Cartwright (1983) asks "Does this law truly describe how bodies behave?" (p. 57) and answers laconically "Assuredly not" (p. 57). She explains further:

> For bodies which are both massive and charged, the law of universal gravitation and Coulomb's law (the law that gives the force between two charges) interact to determine the final force. But neither law by itself truly describes how the bodies behave. No charged objects will behave just as the law of universal gravitation says; and any massive objects will constitute a counterexample to Coulomb's law. *These two laws are not true: worse they are not even approximately true.*
>
> (Cartwright, 1983, p. 57, emphasis added)

A recent appraisal of Newton's contribution provides a better understanding of his methodology:

> Our understanding of … Newton's accomplishment, should therefore acknowledge the increasingly important role played by hypotheses. Thus Whewell urged that we should see Newton's declared rejection of hypothetical reasoning as no more than an expression of a tendency "prevalent in his time", and of his reaction to "the rash and illicit general assumptions of Descartes".
>
> (Gower, 1997, p. 131)

This leads to a dilemma: did Newton formulate his law of gravitation based entirely on experimental observations (quantitative imperative)? If the answer is in the affirmative then he should have been aware that charged bodies would not follow the law of gravitation. Insight from Giere (1999) can help to resolve the dilemma:

> Most of the laws of mechanics as understood by Newton, for example, would have to be understood as containing the proviso that none of the bodies in question is carrying a net charge while moving in a magnetic field. That is not a proviso that Newton himself could possibly have formulated, but it would have to be understood as being regularly invoked by physicists working a century or later.
>
> (Giere, 1999, p. 91)

To recapitulate: it is plausible to suggest that if Newton did not entertain (or was not aware) that charge on a body could influence the force exerted between two bodies then his quantitative data would have led to a different law of gravitation, namely, one that would include the following *ceteris paribus* clause "if there are no forces other than gravitational forces at work" and then the rest of the law as traditionally presented would follow. This clearly shows that in order to formulate his law of gravitation Newton inevitably resorted to idealization based on a hypothesis (despite claims to the contrary), that is an integration between the quantitative imperative (data) and the imperative of presuppositions, facilitating conceptual understanding.

The Chemical Revolution: Did Lavoisier Refute Phlogiston Theory?

The Chemical Revolution in the 18th century associated with A. Lavoisier (1743–1794) provides an interesting episode in the history of science. According to the positivist/inductivist interpretation, before Lavoisier did his experiments scientists accepted G. Stahl's (1660–1734) theory, who sustained that a burning substance emitted phlogiston and hence should decrease in weight. Thus, according to the phlogiston theory, metals on being burnt in air must decrease in weight due to the loss of phlogiston. Lavoisier's experiments in 1772 showed that when phosphorus was burned in a volume of air confined over water, the volume of air was reduced, and the product of combustion weighed more than the original. According to the positivist interpretation, this decisively refuted phlogiston theory. Many chemistry textbooks follow this interpretation (cf. Brown & LeMay, 1987, p. 5 as an illustration). In spite of the popularity of this interpretation, the history of chemistry shows that as early as 1630 (i.e., long before Lavoisier came on the scene) J. Rey (1630) had reported that it was common knowledge that metallic oxides weigh more than the metals from which they were prepared. According to Musgrave (1976), who gives a detailed account of this episode, "if Lavoisier's 1772 experiment refutes phlogiston theory, then phlogiston theory was *born* refuted" (p. 183, original italics). For more details and different interpretations of the Chemical Revolution, see Perrin (1988), Thagard (1990) and Niaz (2008c).

Mendeleev and the Periodic Law

Most of the pioneering work of D. Mendeleev related to the periodic law of chemical elements was conducted from 1869 to 1889, before Thomson (1897), Rutherford (1911), Bohr (1913) and Moseley (1913) laid the foundations of the modern atomic theory. So how could Mendeleev conceptualize periodicity of chemical elements as a function of atomic theory? Despite Mendeleev's own ambivalence and ambiguity, a recent historical reconstruction (Niaz, Rodríguez & Brito, 2004) has demonstrated that Dalton's atomic theory (formulated in the early 19th century) implicitly guided Mendeleev's formulation of the periodic law. Even in his first publication, Mendeleev (1869, p. 405) referred to the relationship, albeit implicitly, between periodicity and the atomic theory by using atomic weights and valence. Just like Newton, Mendeleev (1879) also declared, "I shall not form any hypotheses, either here or further on, to explain the nature of the periodic law" (p. 292). Ten years later in his famous Faraday Lecture, Mendeleev (1889) not only attributed the success of the periodic law to Cannizaro's ideas on atomic theory but went on to explicitly formulate the following hypothesis:

> the veil which conceals the true conception of mass, it nevertheless indicated that the explanation of that conception must be searched for in the masses of atoms; the more so, *as all masses are nothing but aggregations, or additions, of chemical atoms.*
>
> (p. 640, emphasis added)

Indeed, the "veil" always conceals the presuppositions, while letting the quantitative data play the dominant role. Later in the Faraday Lecture, Mendeleev (1889) even attempted to lift the "veil" himself "while connecting by new bonds the theory of the chemical elements with *Dalton's theory of multiple proportions, or atomic structure of bodies, the periodic law opened for natural philosophy a new and wide field for speculation*" (p. 642, emphasis added).

Mendeleev was fully aware that the intellectual milieu of the 19th century required that "scientific theories" be based entirely on experimental evidence (cf. his controversy with the chemist M. Berthelot). The dominant role played by positivism throughout the 19th century has been recognized by philosophers of science (Brush, 1976; Gavroglu, 2000; Holton, 1992). E. Rutherford (1915), an experimentalist *par excellence*, recognized this influence in eloquent terms:

> it is of interest to recall that less than 20 years ago there was a revolt by a limited number of scientific men against the domination of atomic theory in chemistry. The followers of this school considered that the atomic theory should be regarded as a mere hypothesis, which was of necessity unverifiable by direct experiment, and should, therefore, not be employed as a basis of explanation of chemistry … This tendency advanced so far that textbooks were written in which the word atom or molecule was taboo.
>
> (p. 176)

Apparently, Mendeleev's dilemma was that, on the one hand, he could rightly claim that the periodic law was based on experimental observations (quantitative imperative) of chemical elements (an aspiration of scientists in the 19th century), and yet, on the other, he could not give up the bigger challenge, namely, the possible causes of periodicity and hence the role played by the atomic theory (imperative of presuppositions).

Einstein and the Michelson–Morley Experiment

Experimenticism, according to Holton (1969), is the "unquestioned priority assigned to experiments and experimental data in the analysis of how scientists do their own work and how their work is incorporated into the public enterprise of science" (p. 977). The Michelson–Morley experiment of 1887, which provided a "null" result with respect to the ether-drift hypothesis, namely, no observable velocity of the Earth with respect to the aether, is a good example of experimenticism. Many leading scientists and popular textbooks have generally attributed the origin of Einstein's special relativity theory in 1905 to the Michelson–Morley experiment. Details of the experiment and how it had to be repeated many times in order to provide consistent results go beyond the scope of this study. Nevertheless, Einstein himself, despite some ambivalence, was quite forthright in setting the record straight and following is an example:

> In my own development Michelson's result has not had a considerable influence. I even do not remember if I knew of it at all when I wrote my first paper on the subject (1905). The explanation is that I was, for general reasons, firmly convinced that there does not exist absolute motion and my problem was only how this could be reconciled with our knowledge of electro-dynamics. One can therefore understand why in my personal struggle Michelson's experiment played no role or at least no decisive role.
>
> (Einstein's letter to Davenport, February 9, 1954, reproduced in Holton, 1969, p. 969)

This shows clearly that Einstein's primary concern was his presuppositions with respect to special relativity theory and the experimental evidence (Michelson-Morley experiment) was, at best, of secondary importance.

Thomson and the Cathode Rays Experiment

J.J. Thomson (1897) is generally credited to have discovered the electron while doing experiments with respect to the nature of cathode rays. Determination of the mass-to-charge (*m/e*) ratio of the cathode rays can be considered the most important experimental contribution of Thomson (Niaz, 1998). Yet, he was neither the first to do so nor the only experimental physicist. Kaufmann (1897) and Wiechert (1897) also determined the *m/e* ratio of cathode rays in the same year as Thomson and their values agreed with each other. If science is entirely "empirical" based on the quantitative imperative, what made Thomson's work different from that of Kaufmann and Wiechert? Falconer (1987) has explained the difference cogently:

> Kaufmann, an ether theorist, was unable to make anything of his results. Wiechert, while realizing that cathode ray particles were extremely small and universal, lacked Thomson's tendency to speculation. He could not make the bold, unsubstantiated leap, to the idea that particles were constituents of atoms. Thus, while his work might have resolved the cathode ray controversy, he did not "discover the electron".
>
> (p. 251)

The rationale behind the experimental determination of the *m/e* ratio was provided by the imperative of presuppositions.

Millikan–Ehrenhaft Controversy and the Oil Drop Experiment

The oil drop experiment played an important role in the determination of the elementary electrical charge (electrons), and is generally considered to be a classic and "beautiful" experiment (Niaz, 2000a). The oil drop experiment is still difficult to perform (cf. Niaz, 2003) and the acceptance of the elementary electrical charge was preceded by a bitter dispute between R.A. Millikan (1868–1953) and F. Ehrenhaft (1879–1952) that lasted for many years (1910–1925). Both Millikan and Ehrenhaft obtained very similar results and yet Millikan was led to formulate the elementary electrical charge (electrons) and Ehrenhaft to fractional charges (sub-electrons). Holton (1978a) has presented a detailed reconstruction of the research methodologies of Millikan and Ehrenhaft and concluded: "It appeared that the same observational record could be used to demonstrate the plausibility of two diametrically opposite theories, held with great conviction by two well-equipped proponents" (pp. 199–200).

Perl's Search for Fractional Charges (Quarks) and the Philosophy of Speculative Experiments

Despite belief in the elementary electrical charge (Millikan), physicists are still looking for fractional charges (quarks) in various experimental setups, such as accelerators, cosmic rays, bulk matter and an automated Millikan liquid drop method, quite similar to Millikan's original design. Martin Perl (recipient of the 1995 Nobel Prize for Physics) and colleagues at the Stanford Linear Accelerator Center are searching for quarks with a methodology quite similar to that of Millikan (Perl & Lee, 1997). Interestingly, they elaborate on their methodology in terms that would be considered controversial by most practicing physicists and even perhaps many qualitative researchers in education:

> Choices in the design of speculative experiments usually cannot be made simply on the basis of pure reason. The experimenter usually has to base her

> or his decision partly on what feels right, partly on what technology they like, and partly on what aspects of the speculations they like.
>
> (Perl & Lee, 1997, p. 699)

These authors have acknowledged many facets of their controversial methodology which were also used by Millikan, although the latter did not do so in public. Perl and Lee's (1997) philosophy of speculative experiments helps to understand and differentiate cutting-edge from conventional (cf. Kuhn, 1970, normal science) experimental work. Furthermore, it demonstrates that tension between the quantitative imperative and the imperative of presuppositions (philosophy of speculative experiments) continues to be an important aspect of the scientific research methodology.

It is interesting to note that the research methodology of scientists reported in this section spans a period of almost 300 years (late 17th to late 20th century: Newton, Mendeleev, Einstein, Thomson, Rutherford, Millikan, Ehrenhaft, Perl), and still all used some form of Galilean idealization, which shows the role played by the imperative of presuppositions. Interestingly, in order to guide through the confusing terrain of educational research, Phillips (2005b) has suggested a similar historical perspective: "serious misunderstandings of the nature of scientific inquiry underlie some of the recent criticisms of research in education—indeed, Galileo, Harvey, Newton, Pasteur, and Darwin would not be rated as rigorous researchers according to the US [NRC, 2000] criteria" (abstract).

Research Programs in the Physical and Social Sciences as Galilean Idealizations

Historical episodes presented in the previous section show that the image of the physical (hard) sciences generally presented does not agree with its reconstruction in the contemporary philosophy of science. This problem is compounded by the fact that, "social sciences have almost always tried to mimic the so-called hard sciences. We have accepted their paradigms and elevated their ways of knowing even when 'hard scientists' themselves challenge them (Kuhn, 1962)" (Ladson-Billings & Donnor, 2005, p. 288). What the social sciences have tried to "mimic" is precisely the quantitative imperative of the physical sciences. Michell (2003) has traced this influence in psychology to the work of influential psychologists, such as Boring (1929), Eysenck (1973) and Spearman (1937). Niaz (2005a), on the contrary, has shown that, similar to the physical sciences, progress in psychology is also characterized by a confrontation between the quantitative imperative and the imperative of presuppositions, which constitutes an example of the application of the Galilean idealization in the social sciences. As an example, let us consider Piaget's developmental stages within the framework of his genetic epistemology. Brainerd (1978) and others have critiqued Piaget's developmental stages on empirical grounds, namely children and adolescents do not acquire the different stages at the ages stipulated by Piaget, and hence his theory has been falsified. This is a very Popperian approach to understand progress and ignores the fact that Piaget's *oeuvre* is based on the presupposition that developmental stages correspond to an epistemic subject—universal scientific reason, ideally present in all human beings (for details, see Kitchener, 1986, 1987, p. 355; Niaz, 1991a, p. 570). Piaget was not studying the average of all human abilities (quantitative imperative), but rather the ideal conditions (imperative of presuppositions) under which a psychological subject (a particular person) could perhaps attain the competence exemplified by the epistemic subject (cf. Beth & Piaget, 1966, p. 308). This distinction is important as many educational researchers frequently attribute the performance of children on the different Piagetian tasks as a manifestation of the psychological subject, whereas it actually represents the competence of the epistemic subject (cf. Novak, 1977, pp. 464–465).

Piaget's theory of cognitive development has been extended by Pascual-Leone (1970), which constitutes yet another example of the application of the Galilean idealization. Piaget builds a general model by neglecting individual differences (ID), that is, studies the epistemic subject, whereas Pascual-Leone (1987), by incorporating a framework for ID variables (e.g., information processing), studies the metasubject—the psychological organization of the epistemic subject. Niaz (1992) has demonstrated an epistemic transition between Piaget's epistemic subject and Pascual-Leone's metasubject, leading to a theory with greater explanatory power ("progressive problem-shift", cf. Lakatos, 1970).

This shows quite clearly that it was precisely the imperative of presuppositions that led Piaget to postulate the epistemic subject and Pascual-Leone the metasubject, and that progress in psychology has also followed a confrontation between the quantitative data and its interpretation within a theoretical framework. Interestingly, both Piaget and Pascual-Leone's research programs follow the Galilean idealization quite closely, similar to the research programs of Newton, Mendeleev, Einstein, Thomson, Rutherford, Millikan and Perl in the physical sciences. The pioneering work of Pascual-Leone (1970) and Kitchener (1987) to reconstruct cognitive development (similar to the physical sciences) can perhaps be followed in other areas of social science. To sum up, social scientists do not have to "mimic" or emulate the naive falsificationism of scientific theories based on empirical evidence as presented by positivist philosophers of science (and sometimes even by physical scientists themselves), but rather to reconstruct and understand the dynamics of research programs based on Galilean idealizations.

Integration of Qualitative and Quantitative Research Methods

A major argument in favor of mixed methods (integrative) research programs in education is that it provides a rationale for hypotheses/theories/guiding assumptions/presuppositions to compete and provide alternatives. This facilitates the reconstruction of historical episodes in physical science and cognitive psychology based on Galilean idealizations. The historical reconstruction helps to juxtapose empirical evidence in the context of arguments and counter-arguments of the different competing groups of researchers. It is precisely for this reason that

Phillips (2006) has critiqued evidence-based research on the grounds that "good evidence" can be vitiated by being incorporated into a poor or incomplete argument. In the absence of such confrontations and rival interpretations, which may last for many years, it would be extremely difficult to understand the implications of both qualitative and quantitative research. Campbell (1988b) has expressed this in cogent terms, "At any given time, even in the best of science (even in Physics), we are in a historical context and our experiments and our theoretical arguments are historically imbedded" (p. 319). Lakatos' philosophy of science explicitly endorses such strategies in which theories can coexist, whereas Kuhn's philosophy contemplates outright acceptance or rejection of theories. In the context of educational research, Shulman (1986) had foreseen the negative implications of this aspect of Kuhn's philosophy:

> Where Kuhn erred, I believe, is in diagnosing this characteristic of the social sciences as a developmental disability ... it is far more likely that for the social sciences and education, the coexistence of competing schools of thought is a natural and quite mature state.
>
> (p. 5)

Campbell (1988b) goes beyond by emphasizing not only the coexistence of theories but rather the proliferation of "plausible rival hypotheses" and explicitly endorses integration, "*the mistaken belief that quantitative measures* [quantitative imperative] *replace qualitative knowing* [imperative of presuppositions]. Instead, qualitative knowing is absolutely essential as a pre-requisite for quantification in any science" (p. 323, original italics).

Friman, Allen, Kerwin and Larzelere (1993) conducted a citation analysis of the Kuhnian displacement thesis and found that the increasing popularity of cognitive psychology (emergent paradigm) does not necessarily signal the demise of behavioral psychology and psychoanalysis. These authors suggest a more ecumenical alternative,

> such a philosophy should allow for the productive coexistence of alternative approaches to psychology. Two possibilities include the philosophies of Lakatos and Laudan. Both replaced the Kuhnian notion of paradigm with their own concepts. Lakatos proposed "research programs" and Laudan proposed "research traditions".
>
> (p. 664)

Similarly, Newell (1992) has endorsed a similar thesis:

> Theories are always approximate, often deliberately so, in order to deliver useful answers. Theories cumulate, being refined and reformulated, corrected and expanded. This view is Lakatosian, rather than Popperian: A science has investments in its theories and it is better to correct one than to discard it.
>
> (p. 425)

Barker and Gholson (1984), after tracing the dynamics of progress in the psychology of learning, concluded:

> the sort of competition among large-scale units described by Lakatos can readily be identified in the history of the psychology of learning and show that the pattern of historical development in the last 50 years conforms to Lakatos' model and not to Kuhn's.
>
> (p. 228)

These suggestions by Campbell, Shulman, Friman and colleagues, Newell, and Barker and Gholson, provide a blue-print for the integration of qualitative and quantitative research methodologies, within a perspective based on the contemporary philosophy of science.

Methodologists Need to Catch Up with Practicing Researchers

According to Saloman (1991), research methodology in education must adapt to the problem under study. Similarly, Montero (1992), a leading Venezuelan qualitative researcher, has recommended, "The method follows the problem and never vice versa" (p. 68). Given the increasing use of mixed methods in education, and the fact that practicing researchers frequently ignore the recommendations of the methodologists, Johnson and Onwuegbuzie (2004) have concluded, "It is time that methodologists catch up with practicing researchers!" (p. 22). Although, mixed methods research is still in its adolescence, its use and popularity in the social sciences has led to the publication of a *Handbook of Mixed Methods in Social and Behavioral Research*, as "mixed methods research has evolved to the point where it is a separate methodological orientation with its own worldview, vocabulary, and techniques" (Tashakkori & Teddlie, 2003, p. x). Although, this is not a review of mixed methods research, it is important to note the following important developments: (a) new journal, *Journal of Mixed Methods Research* (Sage), started in 2006, which shows the need of an organized community of researchers; (b) publication of a handbook of mixed methods in 2003; and (c) there are many suggestions in the literature pertaining to improvement of mixed methods, of which I mention two: identification of critical appraisal criteria (Sale & Brazil, 2004) and teaching methodology courses that include both quantitative and qualitative techniques, so that the researcher can decide with respect to integration (Onwuegbuzie & Leech, 2005). It is the methodologist who has to catch up with the practicing researcher.

Let us now go back and catch up with the development of the research program of Guba and Lincoln (reported in a previous section, based on Lincoln, 1989). Almost 15 years later Guba and Lincoln (2005) have stated in the *Handbook of Qualitative Research*:

> Positivists and postpositivists alike still occasionally argue that paradigms are, in some ways, commensurable; that is, they can be retrofitted to each

> other in ways that make the simultaneous practice of both possible. We have argued that at the paradigmatic, or philosophical, level, commensurability between positivists and postpositivist worldviews is not possible, but that within each paradigm, mixed methodologies (strategies) may make perfectly good sense.
>
> (p. 200)

Some readers may find a certain change in the theoretical framework of Guba and Lincoln. However, a closer look shows that they still hold on to the rigidity of research paradigms in the Kuhnian sense and that paradigms are incommensurable. In the same handbook, Denzin and Lincoln (2005) have explicitly referred to mixed methods research (especially Teddlie & Tashakkori, 2003) under the heading *Mixed-Methods Experimentalism*, in the following terms:

> Mixed-methods designs are direct descendants of classical experimentalism. They presume a methodological hierarchy in which quantitative methods are at the top and qualitative methods are relegated ... The mixed-methods movement takes qualitative methods out of their natural home, which is within the critical, interpretive framework ... It excludes the stakeholders from dialogue and active participation in the research process. This weakens its democratic and dialogical dimensions and decreases the likelihood that previously silenced voices will be heard.
>
> (p. 9)

This makes interesting reading and raises many issues:

a. classical experimentalism played a role and may still guide research in the future—can any discipline simply break off with its research tradition? Let us pose the dilemma in the following manner: can the followers of quantum mechanics simply deny the research tradition of Newtonian physics? In spite of their differences, Newtonian physics is still considered to be a limiting case of quantum mechanics (cf. Cartwright, 1999, pp. 231–232). Breaking off with research traditions is perhaps a reminder of Kuhnian displacement of paradigms;
b. classrooms in most parts of world are complex "conglomerates of interdependent variables", which may require simplification of the discreet variables (idealization, cf. McMullin, 1985) or a more holistic qualitative approach, depending on the problem to be studied (cf. Saloman, 1991), so that the method follows the problem. Can any particular approach monopolize the research strategies within a single epistemological framework? The answer perhaps lies with the practicing researchers;
c. with respect to a "natural home". Is it possible to assign a particular methodological approach a "home"? Mixed methods researchers have shown that this is precisely not a very democratic alternative;
d. with respect to "active participation" and "silenced voices". These are much cherished objectives but, nevertheless, cannot be an end in themselves.

> Participation and the possibility to voice grievances are a means to an end, namely, the community wants to solve its problems and as researchers we have to first study the problem and then devise the methodology and not vice versa.

Conclusion: Let the Problem Situation Decide the Methodology

Some of the leading qualitative researchers in education (Guba and Lincoln) do not recommend integration of qualitative and quantitative research methodologies. However, given the increasing use and popularity of mixed methods research programs in education, Johnson and Onwuegbuzie (2004) have proposed a framework based on pragmatism (Peirce, James & Dewey). This pragmatism contrasts with the theoretical framework of Guba and Lincoln, which is deeply influenced by the Kuhnian thesis of incommensurability and displacements of paradigms.

This study provides an alternative based on contemporary philosophy of science (Lakatos, Giere, Cartwright & Holton), which emphasizes that most research in the physical sciences has been a transaction between quantitative data (quantitative imperative) and qualitative and controversial interpretations (imperative of presuppositions). Quantitative data (empirical evidence) by itself does not facilitate progress (despite widespread belief to the contrary), neither in the physical sciences nor in education. A historical reconstruction based on contemporary philosophy of science shows that both Piaget and Pascual-Leone's research programs in cognitive psychology follow the Galilean idealization quite closely, similar to the research programs of Newton, Mendeleev, Einstein, Thomson, Rutherford, Millikan and Perl in the physical sciences. This relationship does not imply that researchers in education have to emulate (or mimic) research in the physical sciences. It is concluded that mixed methods research programs (not paradigms) in education can facilitate the construction of robust strategies, provided we let the problem situation (as studied by practicing researchers) decide the methodology.

Next Chapter

Chapter 3 revealed that qualitative research in general is strongly influenced by Kuhn's philosophy of science. This led these researchers to consider the increasing popularity/acceptance of qualitative research as a prerequisite for the replacement of quantitative research. Consequently, Chapter 4 explored alternative approaches to research methodology based on different philosophical perspectives. In other words, qualitative research does not necessarily have to replace quantitative research but rather the two could be considered as rival research programs (methodologies) and even perhaps the two methodologies could be integrated.

Chapter 4

Exploring Alternative Approaches to Methodology in Educational Research*

Introduction

Recent research in education has dealt extensively with the debate over the use of qualitative or quantitative research methodologies (Bereiter, 1994; Eisner, 1992; Gage, 1989; Guba & Lincoln, 1989; Howe, 1988; Lincoln, 1989; Loving, 1997; Martin & Sugarman, 1993; Maxwell, 1992; Niaz, 1997; Phillips, 1987, 1994b; Schrag, 1992). It appears that most qualitative researchers have been inspired by Kuhn's (1970) thesis of paradigm shifts, so that there are multiple realities—all of which are true at the same time (cf. Niaz, 1996a; Phillips, 1983). Guba and Lincoln (1982) have expressed this point of view very cogently: "[Qualitative researchers] focus upon multiple realities that, like the layers of an onion, nest within or complement one another. Each layer provides a different perspective of reality, and none can be considered more 'true' than any other" (p. 57). Qualitative researchers' debt to the Kuhnian conceptualization of progress in science has been acknowledged explicitly by Lincoln (1989):

> As scientists proceed with their work, anomalies occasionally occur ... Sooner or later, however, the anomalies begin to take on the appearance of regularities, and a crisis is precipitated, according to Kuhn. In simplest terms, the crisis revolves about whether the scientific community will reject traditional theory and begin anew to construct theory which accounts for the anomalies, or whether it will remain tied to classical theory.
>
> (pp. 60–61)

This is a fairly standard Kuhnian account of scientific progress, which has been the subject of criticism in philosophy of science (Lakatos & Musgrave, 1970). According to Lakatos (1970), history of science has been characterized by competing research programs. In the context of educational research, according to Shulman (1986):

> Where Kuhn erred, I believe, is in diagnosing this characteristic of the social sciences as a developmental disability ... it is far more likely that for the

* Reproduced with permission from: Niaz, M. (2004a). Exploring alternative approaches to methodology in educational research. *Interchange, 35*(2), 155–184.

> social sciences and education, the coexistence of competing schools of thought is a natural and quite mature state.
>
> (p. 5)

Against the backdrop of this controversy in both philosophy of science and education, it is not surprising that many researchers have suggested that rival methodologies (qualitative and quantitative) would provide a better forum for a productive sharing of research experiences (Campbell, 1988a, 1988b; Friman et al., 1993; Maxwell, 1990a; Niaz, 1997; Saloman, 1991; Shulman, 1986). According to Shulman (1986): "These 'hybrid' designs, which mix experiment with ethnography, multiple regressions with multiple case studies, process-product designs with analyses of student mediation, surveys with personal diaries, are exciting new developments in the study of teaching" (p. 4).

The main objective of this chapter is to provide in-service teachers an opportunity to familiarize with the controversial nature of progress in science (growth of knowledge) and its implications for research methodology in education.

Rationale and Design of the Study

The study is based on 41 in-service teachers who had enrolled for the required course Methodology of Investigation in Education, as part of their Master's degree program. Of the 41 participants, 12 were English as second language teachers, 10 Spanish teachers, 10 mathematics teachers, 5 chemistry teachers, and 4 physics teachers. Participants' ages ranged: 26–31 years = 13, 32–37 years = 18, 38–45 years = 8 and 46–51 years = 2. Twenty-two teachers gave introductory level university courses in their respective areas and the rest were high school teachers. Teaching experienced ranged: 1–5 years = 17, 6–10 years = 12, 11–15 years = 8 and 16–20 years = 4. Almost all participants had seen a methodology course at the undergraduate level, based on texts such as Kerlinger (1975). Few participants had basic knowledge of the work of Popper, Kuhn, Lakatos and other philosophers of science. Some participants had a basic notion of constructivism but it was not grounded in any particular theoretical framework.

Course Content (Reading List)

The course was based on 20 required readings and was subdivided into the following sections:

Epistemology: 1. Montero (1992); 2. Martínez (1993); 3. Campbell (1988b).
Methodology: 4. Campbell (1988c); 5. Erickson (1986); 6. Giroux (1988); 7. Elliot (1985); 8. Taylor and Bogdan (1984); 9. Freire (1970); 10. Maxwell (1992).
Quantitative research: 11. Niaz (1991b); 12. Johnson (1991).
Qualitative research: 13. Brown (1994); 14. Tobin and LaMaster (1995).
Integrating qualitative and quantitative research: 15. Glasson and Lalik (1993); 16. Gaskins (1994); 17. Niaz (1995a).

Methodological and philosophical reflections: 18. Niaz (1996a); 19a. Maxwell (1990a, 1990b); 19b. Lincoln (1990a, 1990b); 20. Saloman (1991).

This reading list shows how the course was designed explicitly not only to incorporate epistemological, philosophical and methodological issues, but also alternative teaching strategies. Precisely, it was this aspect of the course that facilitated exploration of alternative approaches to methodology, referred to as a "hybrid" design by Shulman (1986, p. 4).

Course Organization and Activities

On the first day of class (2 hours) all participants were provided copies of all the readings and a brief introduction to the salient features of the course were discussed. It was emphasized that the course called for active participation. As all the teachers worked in nearby schools and universities, two types of course activities were programmed:

1. Class discussions were planned on Saturdays of the 3rd, 5th and 7th week of the course (3 hours in the morning and 3 in the afternoon). Readings 1–5 were discussed in the first meeting, readings 6–12 in the second meeting and readings 13–20 in the third meeting. Participants were supposed to have studied each of the readings before the respective meetings. Each meeting started with various questions and comments by the participants. The instructor intervened to facilitate the understanding of the issues involved. (Total time devoted to class discussions = 18 hours.)
2. Class presentations were programmed during the 9th and final week of the course (Monday to Saturday, 8 hours daily, total time = 48 hours). On the first day of class participants had formed groups of two or three and selected one of the 20 readings for their presentation. Each of the groups was assigned 2 hours (1 for presentation and the other for interventions). The presenters were supposed to moderate the discussions. The instructor intervened when a deadlock was reached on an issue. It was expected that the participants would present the important aspects of the readings, with the objective of generating critical discussions. All participants prepared PowerPoint overheads for the presentations.

Methodology

After having discussed in class issues relevant to the growth and meaning of knowledge, Initial and Final exams were designed to cover the following issues:

Initial Exam (first session of the 9th week):
Item 1: The problematic nature of "objectivity" and "scientific method";
Item 2: Understanding of a course topic through different philosophical frameworks;

Item 3: How to introduce new ideas in classroom activities;
Item 4: How to grapple with a contradictory aspect of social constructivism (student evaluation).

Final exam (last session of the 9th week):
Item 5: How to select a research methodology when faced with a problem in the classroom;
Item 6: Like other forms of knowledge constructivism needs critical appraisals;
Item 7: The possibility of integrating qualitative and quantitative research methodologies;
Item 8: How critical appraisals show the need for going beyond social constructivism.

This outline of the Initial and Final exams explicitly shows how course methodology was designed in advance in order to provide an opportunity for a continual appraisal of students' own ideas and the new ideas being introduced.

Results

Although this study generated considerable amount of data, this chapter primarily reports results based on participants' responses to the eight items of the Initial and Final exams. Both exams were "open-book" (about 3 hours each) and participants were allowed to consult any material that they felt could be helpful. All responses were classified into categories generated by the response itself. For example, in Item 1 of the Initial exam, a preliminary study of the responses generated 12 categories. Further analyses based on comparisons between categories reduced this number to seven. Categories are not rigid and some responses could have been classified into more than one category. Responses to each of the eight items are then summarized under a general heading, which helps to understand the growth and meaning of knowledge.

Role of "Objectivity" and the "Scientific Method"

This characterization of the growth and meaning of knowledge is based on participants' responses to Item 1 of the Initial exam.

Item 1

According to the new philosophy of science (Feyerabend, Hanson, Kuhn, Lakatos and Laudan) there is no "objectivity" in science and the "scientific method" does not exist. Against this dilemma, if you have to implement a new teaching strategy in your classroom, how would you achieve consensus (validity)?

Participants' responses were classified into seven categories and of these two are presented here.

Social Constructivist (n = 12)

Most of these participants referred to at least some aspects of social constructivism explicitly and the following is an example:

> Participant observation ... provides a dialectical perspective, which is structured horizontally and provides significant learning experiences for the students. My strategy would start with students' anecdotal experiences, which would reveal their previous knowledge.... I would eliminate objective tests and include open questions, which would allow the students to think and reflect. Instead of using behavioral objectives I would try to problematize the course content. This would provide the students the opportunity to formulate their own questions, and decide what, how and why to study. As "objectivity" does not exist, I as the teacher cannot impose a topic of study—students formulate their expectations and as the teacher I can coordinate and facilitate. This of course will create a problem with respect to what the official program stipulates. Activities based on the participation of the students would provide the basis of the validity (consensus) of my strategy.

Integration of Qualitative and Quantitative Methods (n = 11)

All the responses in this category subscribed to some form of constructivism and some even to social constructivism. What, however, makes this category different from the previous is that these responses explicitly postulated the integration of qualitative and quantitative methods. The following is an example:

> The investigator perturbs the problem itself. How can there be objectivity if the object of study is the investigator?... The new philosophy of science wants to avoid the predominance of only one method (hypothetico-deductive), precisely because it is impossible that there could be a single "recipe" for explaining something. In order to achieve consensus (validity) it is important not to be led into provocations when different hypotheses are confronted. The subjective aspect of qualitative research can easily be complemented by using a plurality of methods. I would select different elements of the teaching strategy without prejudice as to their origin, so long as they provide "nutrition" for my students and facilitate my formation as "teacher-investigator".

Discussion

These responses show that participants were aware that problematizing the course content leads to conflicts, such as: objective tests/open question, behavioral objectives/comprehension, official program/students' and teachers' expectations, relationship of what is being constructed to what is already known (established knowledge) and how the classroom environment helps to elaborate method. Furthermore, participants were aware that positivism was criticized for the predominance of the hypothetico-deductive method. This critique has led to the acceptance of a plurality of methods and the proliferation of interpretations that facilitate deeper understanding of a problem.

Alternative Approaches to Growth and Meaning of Knowledge

This characterization of the growth and meaning of knowledge is based on participants' responses to Item 2 of the Initial exam.

Item 2

Development of knowledge is conceptualized by the following authors:

Kuhn: Through scientific revolutions that are characterized by abrupt changes that lead to the displacement of one paradigm by the other.
Lakatos: Through competition between rival research programs (instead of paradigms) and not in the monopoly of one program.
Campbell: Through rivalry between different hypotheses that appear plausible for explaining a phenomenon.
Erickson: Through coexistence of the old and the new paradigms. It is difficult to replace an old paradigm by falsification.

Select a topic from your area of work and explain the development of knowledge (historic context) through any of the four conceptualizations presented above.

Responses were classified into six categories and of these three are presented here.

Conceptualizations Based on the Work of Erickson (n = 13)

The following is an example from a mathematics teacher:

> The development of geometry conforms to the thesis of Erickson, namely, the coexistence of Euclidean geometry with the new paradigm (non-Euclidean) based on the geometry of Riemann, Lobatchevsky and Gauss. The old paradigm held its sway for almost 1500 years until the appearance of the new paradigm in the middle of the 19th century. The new paradigm refuted some of the axioms of Euclidean geometry.... Nevertheless, some of the old postulates of Euclidean geometry survived, which we still use in our daily lives.... This shows that the new and the old paradigms can coexist without entering into conflict.

Conceptualizations Based on the Work of Lakatos (n = 12)

The following is an example from a chemistry teacher:

> Evolution of the different atomic models from Dalton to the quantum mechanical model of the atom can be interpreted as competing and rival

> research programs as conceptualized by Lakatos.... Dalton's model of the atom was based on the hard-core idea that the atom was indivisible.... Thomson's experiments led to the postulation of a model of the atom that was divisible. Thus Thomson's model of the atom had to compete with that of Dalton.... Rutherford based on his alpha-particle experiments was led to the nuclear model, which had to compete with Thomson's model of the atom.... Bohr's model of the atom explained the stability of Rutherford's nuclear atom and thus the two had to compete. Bohr's model of the atom could explain only the spectrum of atoms with one electron. Quantum mechanics presented a rival model of the atom that could explain the spectra of atoms with more than one electron.

Conceptualizations Based on the Work of Kuhn (n = 9)

The following is an example from an English teacher:

> The development of different methodologies in the teaching of English as a second language adapts perfectly to the Kuhnian conceptualization. At first we had the grammatical perspective.... By the end of the 19th century we had a radical change, when this was displaced by the method based on the translation of written texts.... In the 1930s, under the influence of Skinnerian behaviorism, this was displaced by the "oral approach". This approach had its days of glory during the II world war when it was used by the US government for teaching German and Japanese.... During this period the previous approach was thrown in the "garbage dump of history"—to use a Marxist terminology. The "oral approach" or the "Army method" as it came to be known later, reigned absolutely till the 1970s. At this stage as behaviorism was being bashed, the "oral approach" was criticized for being savagely repetitious, and it agonized until it was displaced by the "functional approach" with its emphasis on interactions.

Discussion

Responses to Item 2 showed that participants not only used a particular framework but also provided competing interpretations of the same developments in their area of work. For example, a response very similar to the one presented as Kuhnian (teaching of English as a second language) was considered as Ericksonian by another English teacher. Other examples of multiple interpretations of the same developments were also found. This facilitates debate and controversy with respect to the understanding of a discipline by the teachers.

How to Design a New Teaching Strategy?

This characterization of the growth and meaning of knowledge is based on participants' responses to Item 3 of the Initial exam.

Item 3

Explain the design of a new teaching strategy in your classroom, incorporating the following concepts:

a. Modification/elimination/utilization of behavioral objectives (Readings 3 and 7);
b. Learning cycle (Reading 15);
c. A window into the mind of the students (Reading 14);
d. Cognitive conflict (Reading 17);
e. Teaching for comprehension (Reading 7);
f. Generating theme (Reading 9);
g. Teaching experiment (Reading 17);
h. M-demand of a task (Readings 11 and 12);
i. Metaphor comprehension (Reading 12);
j. Signifier–signified (Reading 13);
k. Constituted–constitutive (Reading 13).

Note: Your presentation must include at least three concepts.

Discussion

All participants developed a teaching strategy based on a combination of concepts that was unique. The number of times that a concept was used in the different strategies was the following: a=13, b=13, c=12, d=18, e=14, f=13, g=5, h=5, i=5, j=10, k=6. This shows that all the concepts were used at least once. Most of the participants used at least three concepts and a few used only one or two or more than three. Some of the salient aspects of participants' strategies are summarized below:

1. Concept (a) referred to behavioral objectives and could have been used in three different ways: (aM) Modification, (aE) Elimination and (aU) Utilization. Of the 13 participants who used concept (a), seven preferred (aM), one (aE) and five (aU). Behavioral objectives were discussed several times, especially with reference to Readings 3 and 7. Most participants were quite critical of behavioral objectives and reported various personal experiences. In spite of this, only one participant explicitly suggested elimination (aE), and stated:

   In actual educational practice a student is considered to have learnt if the expected objectives are observed. Evaluation is based on a student's performance on objectives that are conceived outside the classroom. In other words you "teach" for evaluation. I do not agree with this, as it inhibits students' creative and critical abilities.

   All seven participants who selected modification (aM) of behavioral objectives suggested that the objectives be discussed in class and modified according to the needs of the students. Five participants insisted on the use (aU) of behavioral objectives. Interestingly, at least three of these combined (aU) with other concepts that contradicted the basic idea of behavioral objectives. For example, one of the participants selected using (aU) along with (c) which suggests a window into the students' mind. Another participant selected using (aU) along with (e) which is based on teaching for comprehension. The third participant selected using (aU) along with (k) which emphasizes the distinction between constituted and constitutive nature of students' learning. These results indicate the considerable influence of behavioral objectives in educational practice. One of the participants, who had designed a teaching strategy based on (b), (c), (e) and (h), commented: "Although I have designed a strategy,... and I do want to change my classroom practice, I feel that it is difficult for me to abandon my old positivist conception."

2. Many participants used the different concepts in a somewhat different form with respect to how they were used by the original authors and discussed in class. For example, in the case of learning cycle (b) almost all 13 participants who used it did not explicitly differentiate between the two forms of the learning cycle, namely, the one developed by Karplus et al. (1977) and its later modification (Glasson & Lalik, 1993). It appears that most participants emphasized only the exploration phase, as a means to understand students' prior knowledge of the topic. Some of the participants who used the concept (c), which referred to a window into the students' mind, suggested using it in order to understand the students' prior knowledge. This is not exactly the sense in which the original authors use this metaphor (Tobin & LaMaster, 1995). Participants, by using the original concept in a slightly different sense, enriched their repertoire and hence facilitated the construction of their knowledge.
3. One of the participants, after having described his teaching strategy based on (aU), (d) and (j), concluded on the following optimistic note:

   > With these three concepts we can elaborate a good teaching strategy for the classroom ... and this is only the "tip of the ice-berg", as what lies hidden is even richer—leading to *revolutions*, a la Kuhn, in socio-educative aspects.
   >
   > (original italics)

   It is interesting to note that, on the one hand, this participant maintains the traditional use of behavioral objectives (aU) and, on the other, foresees revolutions in the Kuhnian sense.
4. One of the English teachers presented a strategy based on the following: teaching experiment (g), generating theme (f) and cognitive conflict (d). The generating theme was based on the fidelity of translation from source language (English) to target language (Spanish), using a word-for-word strategy. Students will be presented with, for example, the following statement in English and asked to provide a word-for-word translation: WAR IS A FOUR-LETTER WORD. A word-for-word translation will show to the students that it makes no sense—thus introducing cognitive conflict. Students

are then asked to consider alternatives. Finally, the alternative of translating the context of the statement is suggested, which leads to resolution of the conflict. Further examples are provided. The whole presentation (about two pages) is considered to be a teaching experiment.

5. A physics teacher presented a strategy based on the following: modification of behavioral objectives (aM), cognitive conflict (d) and learning cycle (b). The novel aspect of this strategy is the utilization of the "exploration" phase of the learning cycle for introducing a cognitive conflict. In order to understand the concept of heat, the teacher suggested using the following activity in the laboratory. Students are provided two glasses (A and B) of water at the same temperature. In glass B there is double the amount of water as compared to glass A. Students are asked to introduce a cube of ice (same size) in each of the two glasses. After observing the two glasses for a few minutes, the students are asked: "If the temperature of water in the two glasses is the same, why did the cube in glass B melt first?" The answers given by the students and subsequent interaction provide the opportunity for introducing cognitive conflict. Later the teacher goes on to use the "clarification" and "elaboration" phases of the learning cycle to facilitate students' differentiation between heat energy and temperature.

Should Students Be Evaluated Based on What They Have Learned or What They Should Have Learned?

This characterization of the growth and meaning of knowledge is based on participants' responses to Item 4 of the Initial exam.

Item 4

According to Sarah (Reading 14, p. 238), her classroom experience facilitated the construction of a new metaphor: "a window into the mind of the student". This metaphor implies that the students be evaluated according to what they have learned and not what they should have learned. In contrast, Martha (Reading 15, p. 195) through her experience reached the conclusion that evaluation must include questions that only the very bright students could answer. Considering that both studies were conducted within the framework of "social constructivism": (a) Do you think there is a contradiction in the methodologies of Sarah and Martha? and (b) How would you resolve such a conflict in your classroom?

Discussion

Participants' responses were classified into the following categories.

Perception of No Contradiction and Favor Sarah's Position (n = 6)

These participants did not perceive a contradiction in the classroom practice—social constructivism, of Sarah and Martha (they differed only in their evaluation) and generally favored Sarah's position on evaluation. One of the participants criticized Martha's position on the following grounds: "Martha's system of evaluation will foster competition, leading to 'survival of the fittest' and will force those who are less prepared to improve the quantity of knowledge and not the quality." Another participant expressed a similar concern in the following terms: "We cannot work on the premise that there are some students better prepared than others. This automatically leads to under-estimation of these students and later they will feel oppressed." One of the participants expressed agreement with Sarah's position: "My personal position is inclined towards Sarah. Evaluation is a process that involves value judgment and as such can be flexible, readjusted, corrected and accumulative." One of the participants pointed out that the important thing was not the contradiction between the two positions but rather the fact that:

> Martha never abandoned her initial positivist position, and hence her insistence on evaluating what the students should have learned … in this respect my original position was also marked by the positivist stigma and all that it implies with respect to behaviorism.

Perception of No Contradiction and Favor Martha's Position (n = 5)

Although these participants did not perceive a contradiction between Sarah and Martha's position, they favored Martha's position on student evaluation. One of the participants criticized Sarah:

> We have to adopt Martha's criterion, otherwise we shall be falling into excessive paternalism.… We should not create pedagogical illusions with respect to the fact that all students have the same interest or intellectual disposition for a particular topic. To consider that all students have the same ability for learning is a fallacy and can have unpleasant consequences.

One participant viewed Sarah's position from another critical perspective:

> I am not very convinced of Sarah's "social constructivist" position, as it primarily enabled her to discipline those students who did not allow her to teach … An evaluation based exclusively on what the student has learned will lead to mediocrity.

Another participant suggested: "I would divide the students into groups in such a manner that the better students can interact with others and thus facilitate cognitive conflict leading to improvement in learning."

Perception of No Contradiction and Favor Both Positions (n = 9)

These participants did not perceive a contradiction in the classroom practices of Sarah and Martha and favored both positions. Of the nine participants in this category, three had responded within a social constructivist framework and two had suggested integration of qualitative and quantitative methods on Item 1. These nine participants responded almost evenly on Item 2 with the following distribution: Kuhn = 2, Lakatos = 2, Campbell = 2 and Erickson = 1 (two participants did not subscribe to any conceptualization). Interestingly, of the nine participants, three suggested using behavioral objectives (aU) and another three suggested using a window into the mind of the students (c) on Item 3. These results show that this group of participants tried to reconcile different and opposing views. The following excerpt corroborates this point: "Both Sarah and Martha looked for a framework that adapted to their classroom practice and in spite of the difficulties both were successful."

Perception of a Contradiction and Favor Sarah's Position (n = 7)

These participants explicitly referred to a contradiction between Sarah and Martha's positions and at the same time favored the former. One of the participants criticized Martha: "Martha's methodology is discriminative as only those who learned more could have obtained the maximum grade ... in Sarah's methodology every student is evaluated separately and everyone has the possibility of obtaining the maximum grade." Another participant compared the two methodologies: "Martha's strategy is implicitly based on the positivist (quantitative) philosophy, whereas Sarah's method (qualitative) is in concordance with post-positivist philosophy." One of the participants viewed the contradiction between Sarah and Martha as:

> The dilemma faced by the two is centered in the learning objectives formulated by the teachers.... Instead of including questions that only the better students could answer, I would rather evaluate to what extent my teaching strategies have facilitated learning.

Perception of a Contradiction and Favor Martha's Position (n = 12)

These participants explicitly referred to a contradiction in the strategies used by Sarah and Martha and at the same time favored the latter. One of the participants expressed the following concern: "Given that both methodologies are based on social constructivism, we should accept confrontation as something normal. After all according to constructivism we must contrast different points of view that may lead to alternative solutions." Another participant observed:

> The trouble is that Sarah's major problem was how to discipline the classroom.... In order to achieve that she tried various strategies, until the role of social director within the framework of constructivism allowed her to control the classroom and get the students to learn ... under these circumstances she could not have evaluated beyond what the students had learned.... Martha started with the positivist point of view, which led to dissatisfaction and she adopted social constructivism. This provided the

students greater opportunity to explore, debate, and construct their own ideas.... Thus it is understandable that she included questions that only the better students could answer.

Another participant was quite emphatic in questioning: "How could Sarah have evaluated only what the students had learned?" Another participant questioned Sarah's strategy: "If we have certain educational objectives in the classroom, why we cannot ask the students to demonstrate that they have learned? If the scientists themselves do achieve consensus, why the students cannot understand and interpret such a consensus?" This observation is important given the increasing recognition of the parallel between how scientists work and how students learn, within a history and philosophy of science perspective (cf. Chinn & Brewer, 1993; Duschl, 1990; Matthews, 1994a).

Does the Method Always Follow the Problem?

This characterization of the growth and meaning of knowledge is based on participants' responses to Item 5, which formed part of the Final exam.

Item 5

According to Montero (Reading 1, p. 68): "The method follows the problem and never vice versa" and similarly Saloman (Reading 20) recommends that the research methodology must adapt to the problem under study. Select a problem from your area of study and explain how and why you would use a particular research methodology.

Discussion

Participants presented the Final exam during the last session of the 9th week. This provided an opportunity to observe if the participants had changed or maintained their perspective with respect to what they had already expressed in the Initial exam during the first session of the 9th week. Item 1 of the Initial exam had presented a dilemma to the participants with respect to "objectivity" and the "scientific method". On the other hand, Item 5 of the Final exam presented the participants a possible solution, namely, method always follows the problem. Except for one participant who had some reservations, all the others agreed that method follows the problem. Responses were classified in the following categories.

Qualitative Methodology (n = 19)

The presentation of these participants strongly stressed exclusive use of qualitative methodology. One of the Spanish teachers, who wanted to study students' difficulties in reading and writing, wrote: "I have to use techniques or procedures that allow me to see, listen and observe with care the intonation, pauses, fluidity, etc. This will provide me the data to study this problem." Of the 19 participants in this category, 10 explicitly referred to "participant observation" (Reading 8).

Another Spanish teacher, after agreeing that the method follows the problem, used a novel approach. The teacher faced a dilemma in her high school class: on the one hand the official program requires students to read not only the neo-classical authors but also the contemporary novelists like Mario Vargas Llosa (Peruvian), Gabriel García Márquez (Colombian) and Ernesto Sábato (Argentinian). On the other hand, students do not like to read and write. In addition to this the teacher faced two more problems: "As I work in a small border town [the name is omitted here] I found that students were influenced by the language of the neighboring country. Furthermore, people in general are very superstitious." The teacher decided to combine these elements of the school environment and asked students (in small groups) to relate orally some of their experiences. Later she asked them to write a small story based on their oral presentation. These stories were read, commented on and later corrected in the classroom. At the end of the year the teacher found that the students were interested in reading the works of others. This enabled her to introduce to the students even Martín Fierro (Argentinian), an author who uses lots of metaphors.

Quantitative Methodology (n = 14)

The presentation of these participants stressed exclusive use of quantitative methodology. Of the 14 participants in this category, five based their presentation on control and experimental groups and the use of statistical tests. Some participants considered their presentation to be quantitative as they used standardized tests to diagnose students' difficulties in a particular topic.

Integration of Qualitative and Quantitative Methods (n = 6)

These participants emphasized the use of both qualitative and quantitative methods. Comparison with Item 1 of the Initial exam showed that three participants changed their perspective, and the other three maintained their perspective, by suggesting integration of qualitative and quantitative methods.

Method Does Not Always Follow the Problem (n = 1)

This participant (a Spanish teacher) adopted a fairly novel approach:

> After having studied and reflected upon all the readings and the classroom discussions, I have come to the following conclusions: a) In most of the cases the problem dictates the methodological, epistemological and philosophical nature of the "instruments" to be used; b) In some cases, however, method does not follow the problem, as suggested by Montero. This is because the object of study itself incorporates methodological considerations that serve to study the problem. I do understand that this was an important consideration in the past, but in my opinion it is still in force today.

Critical Appreciation of Constructivism and its Implementation in the Classroom

This characterization of the growth and meaning of knowledge is based on participants' responses to Item 6, which formed part of the Final exam.

Item 6

Evaluate critically the different forms of constructivism. How would you implement constructivism as a methodology in your classroom?

Discussion

Constructivism had formed part of classroom discussions on various opportunities and especially when Readings 13, 14, 15 and 17 were discussed. Most of the participants centered their attention on the differences between radical and social constructivism. Responses were classified in the following categories.

Non-critical Appreciation of Constructivism (n = 13)

These participants highlighted the important aspects of the different forms of constructivism and were in agreement with most of them. An English teacher expressed this consensus: "Both social and radical constructivism constitute two forces with the same perspective that leads the production of knowledge towards the same objective, namely, the search for the *ultimate truth*" (emphasis added). A Spanish teacher emphasized constructivism:

> although it is not easy, constructivism can facilitate conceptual change, i.e., conceptual transformation of the individual based on a horizontal relationship between the teacher and the student—a transition between the cognitive scheme (individual) and a conceptual scheme (group). This form of the educational process is very motivating for me and I felt an empathy with these authors. In my classroom I do use objectives—not in the behavioral sense, but rather in the cognitive sense. I let the students talk about their prior knowledge of a topic, which generally facilitates the elaboration of the objectives. Then I design a series of activities in which students interact and at the end of class the students ask me as to what they should write in their notebooks. I tell them that they had already elaborated the concept and there was little left for the notebook—this is the most satisfying moment for me, as I think I have introduced conceptual change.

Critical Appreciation of Radical Constructivism (n = 21)

These participants generally agreed with social constructivism but were critical of radical constructivism for having emphasized the constitutive nature of knowledge rather than its constituted aspect. Most of them reasoned that by emphasizing the former the role of the teacher in the classroom is neglected (Reading 13). One of the mathematics teachers expressed this dilemma: "If human beings construct their own knowledge of the world in some absolute sense, then it appears that the teacher's participation as a facilitator is neglected."

Critical Appreciation of Social Constructivism (n = 3)

These participants showed a preference for social constructivism as compared to radical constructivism. Nevertheless, they were not satisfied with some aspects, among which evaluation was the most important. All three referred to the dilemmas faced by Sarah (Reading 14) and Martha (Reading 15). One of these participants perceived no contradiction between Sarah and Martha's criteria for evaluation (Item 4 of the Initial exam), whereas the other two did perceive a contradiction. Interestingly, however, all three favored Sarah. One of the participants questioned social constructivism:

> There is a contradiction between what social constructivism advocates with respect to learning for comprehension and evaluation.... It continues to grade and label students, which is of not much help ... we need to go beyond, leading to a culmination which is more consonant with the epistemological bases of social constructivism.... To implement such a methodology is not an easy task. Even in this course we have observed, how we as adults and teachers have "missed" at times the traditional classroom.

Another participant proposed: "I suggest that students be evaluated when they consider themselves to be prepared."

Critical Appreciation of Constructivism (n = 2)

These two participants were critical of all forms of constructivism. One of the participants presented the following thought-provoking critique:

> In my opinion the different forms of constructivism, especially their discussion of constitutiveness and constitutedness of our knowledge—borders on naivete.... This dichotomy does not reflect the reality of the educational process. It worries me to see that, there is a certain messianism centered in the student. It appears as if on contact with us the students will be contaminated.... This perspective does not take into account that both the students and the teachers are victims of the same show-business epistemology (with its ideological background), reinforced by the communications media, which impregnate society with its message.

This message is important, namely, students and the teachers are part of the same society which "oppresses" both (cf. Freire, 1970).

Can We Integrate Qualitative and Quantitative Research Methodologies?

This characterization of the growth and meaning of knowledge is based on participants' responses to Item 7, which formed part of the Final exam.

Item 7

Describe some of the differences between the qualitative and quantitative research methodologies. Do you think that the two methodologies can be integrated?

Discussion

The difference between qualitative and quantitative research methods formed part of classroom discussions on various opportunities and especially when Readings 1, 3, 4, 5, 7, 10, 16, 17, 18, 19 and 20 were discussed. Responses were classified into the following categories.

Integration of Qualitative and Quantitative Methods (n = 33)

After describing qualitative and quantitative methods these participants concluded that the two can be integrated. The following response is quite representative of most participants:

> I think that the emphasis on quantitative methods provides an opportunity to dissociate science from the passionate aspect that all human beings bring to bear on a problem. On the other hand the qualitative methods provide "life" to the "cold figures" that experimental studies tend to produce. It is not desirable that any one form of method may dominate the other,—otherwise it can lead to a monolithic calamity, which is counter-productive in the growth of knowledge.

It is interesting to observe that of the 33 participants in this category, only 11 had advocated integration of qualitative and quantitative methods on Item 1 of the Initial exam, and five had accepted it conditionally. Again, on Item 5 of the Final exam only six participants had accepted integration.

Qualitative Methods Precede Quantitative Methods (n = 7)

These participants emphasized integration in the sense advocated by Campbell (1988a): "*the mistaken belief that quantitative measures replace qualitative knowing*. Instead, qualitative knowing is absolutely essential as a pre-requisite foundation for quantification in any science. Without competence at the qualitative level, one's computer printout is misleading or meaningless" (p. 323, original italics). One of the participants expressed the following opinion: "I think a 'pure' quantitative research does not exist. From its very origin it is impregnated with qualitative knowing. I agree with Campbell that qualitative research precedes quantitative." Of the seven participants in this category, two had previously favored integration of qualitative and quantitative methods on Item 1 of the Initial exam, two had favored a conceptualization of the growth of knowledge based on the work of Campbell on Item 2 of the Initial exam and two had favored integration of qualitative and quantitative methods on Item 5 of the Final exam. These results show a certain change in the thinking of this group of participants.

Integration Presents Methodological, Philosophical and Theoretical Problems (n = 1)

Only one participant expressed a serious concern with respect to integration:

> while supporting the qualitative I do not want to negate the quantitative paradigm—this would lead to the repetition of past errors.... Nevertheless, if integration signifies the fusion of methodological, philosophical, and theoretical elements with all the ideological, religious, and cultural implications, then it seems to be "impossible".

This point of view is important and comes close to that of Lincoln (Reading 19).

Does Social Constructivism Approximate Positivism?

This characterization of the growth and meaning of knowledge is based on participants' responses to Item 8, which formed part of the Final exam.

Item 8

Social constructivism has emphasized the importance of classroom observation for the investigator, through different means (video cameras, tape recorders, interviews, etc.). According to Erickson (1986, Reading 5): "The paradox is that to achieve valid discovery of universals one must stay very close to concrete cases" (p. 130). Similarly, Taylor and Bogdan (1984, Reading 8) have emphasized the importance of *participant observation*. Consider now the following description of positivism: "The world consists of positive and real facts and observable phenomena. Hence scientific knowledge consists only in the description of these phenomena, and the postulation of universal laws." With these considerations in mind, please respond to the following:

a. Knowing that all observation is theory-laden, what advantages or disadvantages do we have in emphasizing the observable and hence the particular?
b. Do you think that the philosophy of social constructivism is very similar to that of positivism?

Discussion

This was the only item that was not based directly on classroom discussions. It deals with important issues being discussed in the literature (cf. Kelly, 1997; Matthews, 1993, 1994a; Nola, 1997; Osborne, 1996; Suchting, 1992; Glasersfeld, 1992). Based on a history and philosophy of science perspective, Matthews (1993) has argued forcefully that at least some forms of constructivism are quite similar to empiricism: "Any epistemology that formulates the problem of knowledge in terms of a subject looking at an object and asking how well what is seen reflects the nature or essence of the object is quintessentially Aristotelian or, more generally, empiricist" (p. 364). The idea of including this item was to see if the participants, after having an overall positive appreciation of constructivism, can evaluate it critically. Responses were classified into the following categories.

Advantages in Observation; Positivism and Social Constructivism Are Very Different (n = 11)

These participants stated categorically on part (a) of the item that observations have advantages, without any mention of disadvantages. On part (b) they stated that positivism and social constructivism were very different. One of the participants justified the advantage of observation in social constructivism in the following terms: "In social constructivism we use 'critical observation', i.e., various observers must coincide in the description [of what is being observed] and its interpretation. On the other hand, positivism uses 'naive observation'." Another participant expressed: "A social constructivist would not use the observations for empirical corroboration of a theory, whereas the positivist does." Another participant explained: "Observations are always subjective.... Nevertheless, a social constructivist interprets the data in its particular context, whereas the positivist seeks observations in order to establish cause-effect relationships or control variables." With respect to part (b) of this item, these participants stated emphatically that positivism and social constructivism were very different. One of the participants expressed the following view:

> The difference lies primarily with respect to constitutedness and constitutiveness of knowledge—positivism subscribes to the former, i.e., knowledge can be imparted as something that has already been constituted, a packet. On the other hand, social constructivism considers knowledge to be constructed by the participation of the students, i.e., constitutiveness of knowledge.

Another participant explained: "In positivism, the observer is independent from the object, whereas in social constructivism he is the participant." Another participant reasoned: "If the two were similar, we would not have so many polemics in the scientific community."

Advantages and Disadvantages in Observation; Positivism and Social Constructivism Are Very Different (n = 8)

These participants stated on part (a) of the item that there were advantages and disadvantages in emphasizing observations, and maintained on part (b) that positivism and social constructivism were very different. Besides mentioning the advantages of observation, these participants emphasized the following disadvantages: neglect of non-observable phenomena, e.g., cognitive abilities and intelligence; lack of generalizations; the researcher may center on something that she/he wants to corroborate—Erickson (1986) refers to this as the deviation toward the typical. On part (b) of the item these participants considered positivism and social constructivism to be very different. One of the participants reasoned: "social constructivism emphasizes comprehension whereas positivism is more concerned about validity. It is precisely for this reason that positivism converted educational practice into a 'recipe'."

No Response with Respect to Observation; Positivism and Social Constructivism Are Different (n = 4)

These participants did not respond to part (a) of the item and in part (b) considered positivism and social constructivism to be different. One of the participants reasoned: "According to the positivists the 'truth' can only be attained by the 'scientific method', whereas the social constructivists hold that it is the practice which determines the problem and hence the method."

Advantages and Disadvantages in Observation; Positivism and Social Constructivism Are Similar (n = 8)

These participants agreed that there are advantages and disadvantages in emphasizing observation (part a) and that positivism and social constructivism were similar (part b). The following excerpt from one of the participants is quite representative:

> Observation provides rich detail of what is being studied ... however, if taken too far it can be counter productive,... it is not difficult to see that there can be other modes of obtaining information ... how can we be so sure of what we can only observe?... how do we counteract the subjectivity of the observer?... With respect to similarity between positivism and social constructivism: I think the two have common features—citations presented in the item convince me. Furthermore, this is not surprising, historically there has to be some vestige of positivism in constructivism.

Disadvantages in Observation; Positivism and Social Constructivism Are Similar (n = 7)

These participants were quite categorical in pointing out the disadvantages of observation (part a) and that positivism and social constructivism were similar (part b). One of the participants pointed out the disadvantages of observation:

> For example, the indiscipline of a student in the classroom does not depend only on the fact that the teacher had not used a dynamic teaching strategy—it can be due to factors that do not belong to the classroom and hence cannot be detected by a simple observation.

Another participant reasoned: "If all observation is theory laden, it is not objective. As a result, the same problem can be viewed differently by different investigators." With respect to the similarity between positivism and social constructivism, one of the participants expressed the view that: "It is not easy to free ourselves of positivism—in spite of all the polemics, it is common to hear people say: 'I have to see in order to believe'."

Conclusion

Classroom discussions, written reports and the written exams helped participants to understand research methodology in the context of alternative approaches. It facilitated the construction of their personal meaning of constructivist methodology and to a certain degree even a critical appreciation. The Initial exam provided them with the following sequence: realization of the dilemma with respect to the scientific method (Item 1); conceptualization of their area of work in different frameworks (Item 2); design of a teaching strategy based on concepts derived from the different perspectives (Item 3); and finally the opportunity to reflect and make a contribution to an issue (evaluation) within the constructivist framework (Item 4). The Final exam in turn: Item 5 provided a possible solution to the dilemma of method by suggesting that the method follows the problem; Item 6 provided the opportunity for a critical evaluation of constructivism; Item 7 suggested an alternative, namely, integration of qualitative and quantitative methodologies; and Item 8 based on recent research explored the possibility of going beyond constructivism.

Some of the important conclusions are presented below:

- Participants were able to understand the basic ideas of constructivist philosophy and its pedagogical implications.
- Participants did not seem entirely convinced that integration of qualitative and quantitative methods was a viable alternative. Participants had three opportunities in this course to select integration as an alternative. On Item 1 of the Initial exam 16 participants accepted it, on Item 5 of the Final exam only six accepted it and on Item 7 of the Final exam 40 accepted it. On Item 7 of the Final exam they were explicitly asked if the two methods could be integrated. No such constraint was present on the two previous items.
- Most participants agreed with Sarah's criteria for evaluation (Tobin & LaMaster, 1995) within the social constructivist framework, namely, students should be evaluated on what they have learned and not what they should have learned. Item 4 of the Initial exam provided participants with an opportunity to evaluate critically this aspect of constructivist methodology. Most of the participants did not perceive a contradiction in Sarah and

Martha's (Glasson & Lalik, 1993) criteria for evaluation and few supported Martha's position. It seems that participants responded consistently as social constructivism provided more and better answers than positivism to their classroom problems. Perhaps constructivism was still quite novel and useful for most participants and hence there was little need to go beyond and explore alternatives.

- At times participants used some of the concepts derived from the different perspectives in a sense different from that of the original authors. It is suggested that such modifications represent creative adaptations of the original concepts.
- It seems that most of the participants were reluctant to accept constructivism as a form of positivism (Matthews, 1994a), a thesis that is gaining support. Item 8 of the Final exam provided participants an opportunity for evaluating this aspect of constructivism. Apparently, only 15 participants accepted with some reservations that constructivism has some vestiges of positivism.

Given the importance of alternative approaches to growth and meaning of knowledge, it is important that teachers be aware of conflicting situations in the classroom that refer to "objectivity", "scientific method", qualitative-quantitative methods, relationship between method and problem, evaluation and a critical appreciation of constructivism. Finally, it is important to note that participants' (in-service teachers) responses to the different items bear witness to their concern, diligence and a spirit of intellectual engagement with the issues involved. This aspect is important for any course that attempts to explore alternative approaches to methodology in educational research.

Next Chapter

After having recognized the importance of alternative approaches (qualitative, quantitative, integrative) to methodology in educational research, Chapter 5 deals with the controversial issue of generalizability in qualitative research. Most qualitative researchers do not accept generalizability as these studies are not based on randomization of samples and application of statistical techniques. It is argued that reporting of data/results in both qualitative and quantitative research inevitably involves interpretation, which often leads to controversies, and hence the degree of generalizability cannot be predicted in advance but rather is determined by a community of researchers. An analogy is drawn with respect to Piaget's methodology, namely it was not based on random samples or statistical treatments and still his work has been generalized (criticisms notwithstanding) in both the psychology and educational literature.

Chapter 5

Can Findings of Qualitative Research in Education be Generalized?*

Introduction

It is generally accepted by qualitative researchers that generalizability is neither desirable nor necessary; as such, studies are not designed to allow systematic generalizations to some wider population. Freidson (1975) has expressed this dilemma in cogent terms: "There is more to truth or validity than statistical representativeness" (p. 272). In a similar vein Guba and Lincoln (1989) have maintained, "Generalization, in the conventional paradigm, is absolute, at least when conditions for randomization and sampling are met" (p. 241). In spite of the difficulties of generalization in qualitative research, Maxwell (1992) has clarified that, "This is not to argue that issues of sampling, representativeness, and generalizability are unimportant in qualitative research" (p. 293). Maxwell then goes on to differentiate between: (a) internal generalizability, namely, generalizing within the community, group or institution studied to persons, events and settings that were not directly observed or interviewed; and (b) external generalizability, namely, generalizing to other communities, groups or institutions. Nevertheless, Maxwell (1992) concedes that this differentiation of generalizability (internal/external) is not always helpful:

> A researcher studying a school, for example, can rarely visit every classroom, or even gain information about these classrooms by other means, and the issue of whether to consider the generalizability of the account for those unstudied classrooms internal or external is moot.
>
> (p. 293)

This, indeed, presents a dilemma for almost all researchers and teachers and has been the subject of considerable controversy in the literature (Denzin & Lincoln, 2000; Eisner, 1997, 1999; Eisner & Peshkin, 1990; Erickson & Gutierrez, 2002; Guba & Lincoln, 1994; Husén, 1997; Knapp, 1999; Macbeth, 1998; Miller, Nelson & Moore, 1998; Niaz, 1997, 2004a; Peshkin, 2000).

It appears that most qualitative researchers consider generalizability to be possible in quantitative research primarily due to the randomization of samples

* Reproduced with permission from: Niaz, M. (2007). Can findings of qualitative research in education be generalized? *Quality & Quantity: International Journal of Methodology, 41*, 429–445.

and the application of statistical techniques. In contrast, this study is based on the premise that reporting of data/results from both qualitative and quantitative research inevitably involves interpretation, which often leads to controversies, and hence the degree of generalizability cannot be predicted in advance but rather is determined by various complex variables. Philosophers of science have referred to this as the under-determination of theory by evidence (cf. Duhem–Quine thesis; Kuhn, 1970; Lakatos, 1970; Quine, 1953). Progress in science has been witness to many such controversies, in which the same data has been interpreted differently by two scientists (cf. Holton, 1978a, 1978b; Niaz, 2000a, for one episode in the history of science). Philosopher–physical chemist Michael Polanyi (1964) has expressed similar ideas eloquently:

> Our vision of reality, to which our sense of scientific beauty responds, must suggest to us the kind of questions that it should be reasonable and interesting to explore. It should recommend the kind of conceptions and empirical relations that are intrinsically plausible and which should therefore be upheld, even when some evidence seems to contradict them, and tell us also, on the other hand, what empirical connections to reject as specious, even though there is evidence for them—evidence that we may as yet be unable to account for on any other assumptions.
>
> (p. 135)

This shows how the plausibility of some conceptions (theory/generalization) can be upheld even if there is evidence to the contrary. In other words, a community of researchers may decide to interpret empirical evidence in a particular way, even if there are alternative interpretations available. This suggests that even in qualitative research generalizability is possible, provided we are willing to grant that our conceptions/theories are not entirely grounded on experimental evidence but rather on the degree to which the community can uphold such a consensus. Campbell (1988b) has expressed this concern for objectivity on the one hand and the process of peer interaction on the other in cogent terms:

> The objectivity of physical science does not come from turning over the running of experiments to people who could not care less about the outcome, nor from having a separate staff to read the meters. It comes from a social process that can be called competitive cross-validation … and from the fact that there are many independent decision makers capable of rerunning an experiment, at least in a theoretically essential form.
>
> (p. 324)

This shows that in order to be scientific (in both social and the natural sciences), objectivity in the handling of data and statistical treatments do not necessarily provide a procedure (scientific method) that infallibly leads to research findings.

As an illustration of how qualitative research has been generalized extensively let us consider Piaget's work at the Center of Genetic Epistemology (Geneva). Most of the studies were based on very small samples with virtually no statistical treatments and represent findings that were not repeated many times before being generalized. Vuyk (1981) has succinctly summarized Piaget's methodology, "Piaget asks a very general epistemological question and then tries to translate it into an experimental situation" (p. 461). The degree to which Piaget's work has been generalized in science education bears witness to its acceptance (a critical appraisal notwithstanding) by the scientific community (cf. Adey & Shayer, 1994; Eylon & Linn, 1988; Lawson, 1985; Shayer & Adey, 1981). Piaget's *oeuvre* has also been the subject of considerable controversy in the psychology literature. Nevertheless, a review of the work of some of the most important critics shows that what was in discussion was not the lack of randomization of his samples, nor adequate statistical techniques. According to Brainerd (1978), an important critic, Piaget's stages of cognitive development are perfectly acceptable as descriptions of behavior, but have no status as explanatory constructs (p. 180). Carey (1986) approves of most of Piaget's descriptive work and the argument seems to center on Piaget's (1985) interpretations, within his metatheory, "It is only when Piaget sought to further explain the differences between young children and adults in terms of domain-general limitations on the child's representational or computational abilities that his interpretations have come under fire" (p. 1129). As to the difference between "descriptions" and "explanations", Pascual-Leone (1988) interprets the problem in an historical perspective by pointing out that what was a good explanation for a theory in the past often becomes a description for the theory superseding it. Niaz (1992) has demonstrated a "progressive problemshift" (Lakatos, 1970), that is, increase in explanatory power in Pascual-Leone's metasubject with respect to the epistemic subject in Piaget's framework. This illustration from Piaget's genetic epistemology shows how a researcher's insight and acumen can facilitate generalizations that are later corroborated (even extended) by the research community.

The objective of this study is to explore the degree to which in-service teachers understand the controversial aspects of generalization in both qualitative and quantitative educational research and as to how this can facilitate problems faced by the teachers in the classroom.

Rationale and Design of the Study

This study is based on 83 in-service teachers who had enrolled for the required course Methodology of Investigation in Education, as part of their Master's degree program, at a major university in Latin America. Of the 83 participants, 26 were Spanish teachers, 21 English as a second language, 17 chemistry, 9 mathematics and 10 physics teachers. Age of the participants ranged from 25 to 45 years and teaching experience ranged from 5 to 15 years (female = 54, male = 29; high school teachers = 55, university teachers = 28). Participants were divided into three groups (1, 2 and 3), depending on their place of work and residence, and the same procedure was followed in all three. Almost all participants had seen a methodology course at the undergraduate level, based on texts such as Kerlinger and Lee (2002). The main objective of the course was to go beyond and provide an opportunity to in-service teachers to familiarize themselves with the controversial nature of progress in science (growth of knowledge) and its implications for research in education. Few participants had basic knowledge of the work of Popper, Kuhn, Lakatos and other philosophers of science. Some participants had a basic understanding of constructivism but it was not grounded in any particular theoretical framework. Readings 6, 7 and 8 led to considerable discussion with respect to social and other forms of constructivism.

Course Content (Reading list)

The course was based on 11 required readings and was subdivided into the following sections:

Epistemology: 1. Montero (1992); 2. Martínez (1993);
Methodology: 3. Erickson (1986); 4. Martínez (1998); 5. Maxwell (1992);
Qualitative research: 6. Brown (1994); 7. Tobin and LaMaster (1995);
Integrating qualitative and quantitative research: 8. Glasson and Lalik (1993); 9. Cortéz and Niaz (1999).
Methodological and philosophical reflections: 10. Niaz (1997); 11. Lincoln and Guba (2000).

The reading list shows how the course was designed explicitly not only to incorporate epistemological, philosophical and methodological issues, but also alternative teaching strategies and issues such as different forms of constructivism and generalizability of research findings. This aspect of the course facilitated exploration of alternative approaches to methodology, referred to as a "hybrid" design by Shulman (1986, p. 4).

Course Organization and Activities

On the first day of class (2 hours) all participants were provided copies of all the readings and the salient features of the course were discussed. It was emphasized that the course called for active participation. As all the teachers worked in nearby schools and universities, two types of course activities were programmed:

1. Class discussions were planned on Saturdays of the 3rd, 5th and 7th week of the course (3 hours in the morning and 3 in the afternoon). Readings 1–4 were discussed in the first meeting, readings 5–8 in the second meeting and readings 9–11 in the third meeting. Participants were supposed to have studied each of the readings before the respective meetings. Each meeting started off with various questions and comments by the participants. The instructor intervened to facilitate understanding of the issues involved. (Total time devoted to class discussions = 18 hours.)
2. Class presentations were programmed during the 10th and the final week of the course (Monday to Saturday, total time = 44 hours). On the first day of class participants had formed groups of three or four and selected one of the 11 readings for their presentation. Each of the groups was assigned 2 hours

(1 for presentation and the other for interventions and discussions). The presenters were supposed to moderate the discussions. The instructor intervened when a deadlock was reached on an issue. It was expected that the participants would present the important aspects of the readings, with the objective of generating critical discussions. All groups prepared overheads based on PowerPoint presentations.

Evaluation

All participants presented an Initial exam in the first session of the 10th week and a final exam during the last session of the 10th week. Both exams were "open-book" (about 3 hours each) and participants were allowed to consult any material that they felt could be helpful. This study is based on participants' responses to Item 1 of the final exam, which consisted of the following research questions:

> Consider the following questions with respect to research in education and you are asked to respond by indicating: (i) In agreement; (ii) In partial agreement; and (iii) In disagreement:
>
> a. Qualitative research as a method can help us to solve all the problems in education.
> b. Quantitative data cannot be used and much less its statistical treatment.
> c. It is not necessary to establish the validity of research but rather its "authenticity" or in other words interpretative validity.
> d. It is neither desirable nor necessary to generalize from results obtained in qualitative studies.
> e. If the phenomena are observed directly (participant observation) greater is the "authenticity" or validity of the research.
> f. Formulation of hypotheses, manipulation of variables, and the quest for causal relations do not help to solve problems.
>
> *Note*: Please justify your response in each case.

This study deals with participants' responses to research question (d), which is considered as:

Research Question I

Consider the following statement with respect to research in education and you are asked to respond by indicating: (i) In agreement; (ii) In partial agreement; (iii) In disagreement:

> It is neither desirable nor necessary to generalize from results obtained in qualitative studies.

Research Question 2

This research question was based on Item 3 of the Final exam and stated:

> Do you think that Sarah's (Reading 7) experience based on social constructivism can be applied in the context of Venezuelan education? If your response is in the affirmative, would this require some changes? If your response is in the negative, please explain.

Note: Group 1 participants responded with respect to Sarah's experience. Group 2 participants responded with respect to Sarah and Martha's experience and Group 3 participants responded with respect to Martha's (Reading 8) experience.

Rationale for Research Questions 1 and 2

The objective of Research Question 1 was to make participants aware of the controversial nature of generalization in educational research. Erickson (Reading 3), Martínez (Reading 4), Maxwell (Reading 5) and Lincoln and Guba (Reading 11) explicitly deal with this issue and provide reasons, arguments and justifications for and against the use of generalization in educational research. On the other hand, the objective of Research Question 2 was to observe participants' responses to a contradictory/conflicting situation, based on actual classroom practice in which two teachers (Sarah and Martha, in the USA), based on their social constructivist framework, adopted different teaching strategies. Sarah, a teacher in Reading 7 (Tobin & LaMaster, 1995), facilitated the construction of a new metaphor, "a window into the mind of the student". This metaphor implies that the students be evaluated according to what they have learned and not what they should have learned. In contrast, Martha, a teacher in Reading 8 (Glasson & Lalik, 1993), through her experience reached the conclusion that evaluation must include questions that only the very bright students could answer. During classroom discussions it was observed that some teachers favored Sarah's strategy, whereas others favored Martha's. Based on this, Research Question 2 provided three alternatives: Group 1 was asked to consider generalization based on Sarah's experience, Group 2 based on both Sarah and Martha and Group 3 based on Martha. The idea behind this was to see if a particular experience (Sarah/Martha) had an incidence on the degree to which participants supported generalization. It was expected that in order to be consistent, those participants who would agree with Research Question 1 would preferably disagree with Research Question 2, or vice versa.

Results and Discussion

Table 5.1 shows that 71% (Categories 1, 3 and 5) of the participants agreed to Research Question 2, which means that the experiences of Sarah and Martha based on social constructivism could be generalized in the context of Venezuelan education. Difference in the performance of Groups 1, 2 and 3 was statistically

Table 5.1 Categorization of Participants' Responses to Research Questions in the Study (n = 83)

Response to			*Number of Participants*			
Category No.	*Research Question 1*	*Research Question 2*	*Group 1 (n = 23)*	*Group 2 (n = 27)*	*Group 3 (n = 33)*	*Total (%)*
1.	Agreed	Agreed	7	11	16	34 (41%)
2.	Agreed	Disagreed	2	–	–	2 (2%)
3.	Partially agreed	Agreed	3	9	6	18 (22%)
4.	Partially agreed	Disagreed	1	1	2	4 (5%)
5.	Disagreed	Agreed	8	6	9	23 (28%)
6.	Disagreed	Disagreed	2	–	–	2 (2%)

Notes
Group 1 participants were asked to generalize based on Sarah's experience (Reading 7).
Group 2 participants were asked to generalize based on Sarah and Martha's experiences.
Group 3 participants were asked to generalize based on Martha's experience (Reading 8).
Categories 1 and 6 were considered Inconsistent, 2 and 5 Consistent, 3 Partially Inconsistent and 4 Partially Consistent.

Research Question 1
Consider the following question with respect to research in education and you are asked to respond by indicating: (i) In agreement; (ii) In partial agreement; (iii) In disagreement.
It is neither desirable nor necessary to generalize from results obtained in qualitative studies.

Research Question 2
Do you think that Sarah's (Reading 7) experience based on social constructivism can be applied in the context of Venezuelan education? If your response is in the affirmative, would this require some changes. If your response is in the negative, please explain.

(Chi Square) not significant, which means that all three groups accepted generalization to about the same extent. This finding is interesting as during class discussions participants became fully aware of the fact that qualitative researchers (Erickson, 1986; Lincoln & Guba, 2000; Maxwell, 1992) do not recommend external generalization. Furthermore, 41% of the participants agreed to both Research Questions 1 and 2 (an inconsistent strategy) and 22% partially agreed to Research Question 1 and agreed to Research Question 2 (a partially inconsistent strategy). This shows that 63% of the participants used a fairly inconsistent approach, namely, agreed/partially agreed that research experience in qualitative studies cannot be generalized and still agreed that the research experience of Sarah and Martha (Research Question 2) can be generalized in the context of Venezuelan education. Interestingly, most participants were not cognizant that their approach was contradictory or at least inconsistent. On the other hand, only 28% (Category 5) of the participants were consistent, namely, disagreed with Research Question 1 and agreed with Research Question 2. How do we explain these findings? In this respect it would be helpful to consult the written justifications/reasons provided by the participants for their selected responses.

Examples of Category I Responses

Participants in this category agreed to both Research Questions 1 and 2 (see Table 5.1), and the following are some of the examples:

First Example

Response to Research Question 1:

> I agree ... What I want to express is illustrated by the following example: Different sculptors can elaborate different sculptures from the same block of marble. This means that different researchers can obtain different results on studying the same problem ... and hence it is not necessary to generalize.

Response to Research Question 2:

> Yes, Martha's experience can be applied in Venezuelan education ... Most of our teachers are authoritative and inhibit students, and this new experience will help to make the teachers: Guides ... critical investigators ... facilitators of a reciprocal learning relationship.

Second Example

Response to Research Question 1:

> I agree, as there are no absolute truths ... and truths are relative and can change. What is a scientific revolution today will not be so tomorrow ... research findings depend on the moment ... in which the study is conducted.

Response to Research Question 2:

> In my opinion, Martha's experience can be applied in the Venezuelan context depending on certain changes that would help the teacher to implement new strategies in the classroom … [Rest of the response describes the different phases of the learning cycle, from Reading 8].

Third Example

Response to Research Question 1: "I agree, in qualitative research the same situation can lead to different results."

Response to Research Question 2: "I think that Martha's experience can be applied not only in Venezuela but in any part of the world."

Fourth Example

Response to Research Question 1: "I agree. In qualitative research generalization (internal) is possible through the development of a theory that not only interprets persons and situations under study but also shows that the same process in different situations can lead to different results."

Response to Research Question 2:

> Yes, I agree that Sarah and Martha's experiences should be applied in the Venezuelan educational context. Both of them were more than willing to recognize their deficiencies and accept changes in their original metaphors related to educational practice … [a succinct summary of the main features of Sarah and Martha's teaching strategies is provided] … If Venezuelan teachers could look into the mirror of Sarah and Martha and accept that we have deficiencies … this could help to facilitate critical learning.

Fifth Example

Response to Research Question 1: "I agree … qualitative studies are usually not designed to permit systematic generalizations to some other population."

Response to Research Question 2:

> Yes these can be applied in the context of Venezuelan education. I admire the decisions of Sarah and Martha, who had well defined ideologies and still decided to change their role in order to facilitate learning in their classrooms … This was a valiant attitude that requires courage.

Sixth Example

Response to Research Question 1: "I agree, and this is because we live in a changing world … to generalize is risky as social situations and contexts may seem to be similar but are not the same."

Response to Research Question 2:

> Sarah and Martha's experiences can be applied in our context ... however, some changes are required, e.g., economic situation of the teachers will have to be improved so as to make working conditions more stimulating, teachers' work load will have to be decreased so that they have more time for preparation of classes, etc.

These examples show that participants in Category 1, while responding to Research Question 1, were quite convinced that findings from qualitative research cannot be generalized as research is a creative process (example about carving sculptures), there are no absolute truths, scientific revolutions (Kuhn) make transfer of knowledge difficult, qualitative research develops theoretical constructs for particular situations, and that generalization involves risks. Interestingly, however, in Research Question 2 these participants had no difficulty in recommending that findings from two qualitative studies (based on Sarah and Martha's experience in the USA) can be applied to a different social context (Venezuela). Most of the participants had no reservations with respect to this generalizability and mentioned various advantages, such as facilitation of a reciprocal learning relationship, a critical appraisal by teachers of their own deficiencies and the courage to change one's convictions (frameworks) in the face of new contradictory evidence. Most teachers also suggested various changes in the present educational system in Venezuela in order to implement Sarah and Martha's teaching strategies, such as better working conditions.

Examples of Category 3 Responses

Participants in this category partially agreed with Research Question 1 and agreed with Research Question 2 (see Table 5.1) and following are some of the examples:

First Example

Response to Research Question 1: "Before deciding whether it is feasible to generalize or not it is important to know the degree to which the phenomenon of interest varies in the surrounding or new context."

Response to Research Question 2:

> I agree that it is feasible to apply the experiences of Martha and Sarah in our context ... However, I do not share the contradiction with respect to evaluation ... Let me explain: If the constructivists state that all levels of formation [experience] are viable then Martha's considerations are irrelevant. This aspect is of concern to me as I agree with Martha that students must be evaluated based on what they should have learned, in order to facilitate greater understanding.

Second Example

Response to Research Question 1: "Generalization would be possible if the same conditions are reproduced in other contexts to be investigated."

Response to Research Question 2: "Yes I agree, if the changes are the product of reflection and paradigmatic controversies (Lakatos). Furthermore, these experiences would help us not only to understand our educational practice but also to construct our own metaphors."

Third Example

Response to Research Question 1:

> I agree partially, as in my opinion even in quantitative research a sample can never be truly representative ... in most studies the sample is "intentional" to a certain extent and depends on the reality and the convenience of the researcher. Hence the criterion for representative samples is not satisfied in either qualitative nor quantitative research.

Response to Research Question 2: "Yes I agree, provided appropriate changes are made and of these the most important is with respect to the formation of the teachers."

Responses in this category attempt to create new interpretations by not strictly following the framework of some of the leading qualitative researchers (Erickson, 1986; Lincoln & Guba, 2000). Participants were willing to look at the degree to which the surrounding context differed from the original before deciding with respect to generalization. Furthermore, it was pointed out that even in quantitative research it is very difficult to work with truly representative samples and most studies are generally based on "convenient" samples (third example). In science education it is frequently observed (cf. Lawson, 1985) that quantitative researchers generally conduct their studies at the institution where they work, and their findings can at best be generalized only in that particular region/country. However, a review of the literature shows that these very regional studies are then generalized in almost all parts of the world (cf. Eylon & Linn, 1988). More recently, Keeves and Adams (1997) have provided support for such sampling procedures:

> the practice has emerged, largely for administrative convenience in testing in order to avoid the splitting of class groups, of drawing a stratified random sample of schools with a probability proportional to size and *with the school as the primary sampling unit.*
>
> (p. 36, emphasis added)

Another important aspect in these responses (second example) is that generalization of research experiences can lead to controversies (an important characteristic of progress in science, cf. Lakatos, 1970) and even facilitate the construction of "our own metaphors" in the new context.

Example of Category 4 Response

Participants in this category partially agreed with Research Question 1 and disagreed with Research Question 2 (see Table 5.1) and the following is an example:

Example

Response to Research Question 1:

> If the original study clearly outlines the context in which a qualitative study was conducted and spells out the recommendations, then it is possible that some aspects could be applied in a different scenario—without, of course leading to a complete generalization.

Response to Research Question 2:

> Sarah's experience left many "blind alleys" and hence I do not think it can be applied in our context. Her fundamental problem was to maintain discipline in the class ... for this purpose she had to call on the presence of one of the school Directors ... this is very common in Venezuela. Based on social constructivism Sarah evaluated students based on what they had learned and they improved their grades ... this is also quite frequent in Venezuela, so that both the students and the parents are satisfied. This ignores the question of students' preparation for future studies. It seems that Sarah is doing what many teachers in Venezuela do anyway, that is lowering the quality of education.

Responses in this category partially agreed that some aspects of qualitative research can be generalized. However, on considering Research Question 2 these participants found some aspects unacceptable and thus disagreed (e.g., blind alleys in Sarah's experience with respect to discipline and evaluation).

Examples of Category 5 Responses

Participants in this category disagreed with Research Question 1 and agreed with Research Question 2 (see Table 5.1), and the following are some of the examples:

First Example

Response to Research Question 1:

> Once qualitative research has been conducted one must think about the transferability of particular aspects in order to be incorporated in a universal perception. Generalization, is thus, a necessity so as to understand that the universal is not that which repeats frequently (e.g., Piaget's methodology), but rather what belongs to the essence of what is being studied. This is

difficult even in quantitative research as often the samples are not totally representative, due to the difficulties associated with the collection of large amounts of data.

Response to Research Question 2: "I agree with Sarah's experience in order to facilitate creative capacity for understanding new concepts."

Second Example

Response to Research Question 1: "It is possible to generalize provided we are willing to recognize that there will always be a margin of error/discrepancy between the original situation and the one to which it is being applied."

Response to Research Question 2: "Martha's experience can be applied in our context provided the teacher is convinced that it will facilitate significant improvement in learning."

Third Example

Response to Research Question 1: "Generalization is difficult in qualitative research. However, what is the alternative—should we dedicate all our lives to study the whole population on every problem of interest? Hence I disagree."

Response to Research Question 2: "Without suggesting that the learning cycle (used by Martha) is a panacea, as theories change, I think that it can be applied in our context."

Participants in this category were consistent by disagreeing with Research Question 1 and agreeing with Research Question 2 (i.e., accepted generalization). In a sense these responses innovated by not following the strictures of most qualitative researchers. The main argument seems to be that although transferability of findings from one context to another is not automatic, it still remains a necessity. Many participants referred to Piaget's methodology (which was discussed in class) as a prime example of how qualitative research can be generalized. Furthermore, it was recognized that there will always be discrepancies between the original and the applied contexts, and this precisely leads to controversies and not panaceas, and hence progress in educational research.

Conclusion

In-service teachers in this study participated in a Methodology of Investigation in Education course in which various aspects of qualitative and quantitative research were discussed based on a philosophy of science perspective. This facilitated the understanding of the controversial nature of growth of knowledge and its implications for research in education. A major finding of this study is that after having studied and discussed the role of generalization in qualitative and quantitative research, almost 91% of the teachers agreed that generalization (external) in a different social context is feasible. Furthermore, almost 63% of the participants used a fairly inconsistent approach, namely, agreed/partially agreed with Research Question 1 (Categories 1 and 3), that is, research experience in qualitative studies cannot be generalized (external) and still agreed that the research experience of Sarah and Martha in the USA (Research Question 2) can be generalized in the context of Venezuelan education. Interestingly, almost 28% of the participants were consistent (Category 5) in first accepting the possibility of generalization in qualitative research and later its external application.

It is important to note that teachers were fully aware as to why qualitative researchers (Erickson, 1986; Lincoln & Guba, 2000) do not recommend generalization and still agreed to external generalization in a different social context based on the following reasons/justifications: (a) facilitation of reciprocal learning relationship; (b) a critical appraisal by teachers of their own deficiencies; (c) courage to change one's convictions (frameworks) in the face of new contradictory evidence; (d) even in quantitative research, "convenient" rather than random samples are used. Apparently, participants argued that if this was one of the major arguments of qualitative researchers for not generalizing then it could be circumvented; (e) generalization can lead to controversies (an important characteristic of progress in science) and even facilitate construction of "our own metaphors" in the new context; (f) generalizability from one context to another is not automatic but still remains a necessity (Piaget's methodology was used as a prime example); (g) there will always be discrepancies between the original and the applied contexts, and this precisely leads to controversies and not panaceas, and hence progress in educational research.

A major argument of qualitative researchers for not generalizing from qualitative studies is that this research is not based on sufficiently representative samples and adequate statistical controls. In this context, it is pertinent to ask whether most of Piaget's work was based on representative samples. A review of the literature shows that this was not the case. So how did Piaget's work come to be generalized and accepted by the educational research community? A plausible reason is provided by Piaget's differentiation between the epistemic and the psychological subjects and that his genetic epistemology primarily referred to the former. Epistemic subject refers to the underlying rationality (universal scientific reason) ideally present in all human beings (for details, see Kitchener, 1986, p. 81, 1993; Niaz, 1991a, p. 570). In other words, Piaget was not studying the average of all human abilities (hence lack of statistical treatments and random samples), but rather the ideal conditions under which a psychological subject (a particular person) could perhaps attain the competence exemplified by the epistemic subject.

Finally, it is important to note that recent research has recognized the need for integrating research methodologies, that is, utilize both quantitative and qualitative techniques when conducting educational research (Johnson & Onwuegbuzie, 2004; Niaz, 1997, 2008d; Onwuegbuzie & Leech, 2005). This provides further warrant for generalization of findings of qualitative research in education.

Next Chapter

After having recognized the importance and feasibility of generalization in both qualitative and quantitative educational research, Chapter 6 deals with (among other issues) the difference between validity in quantitative research and authenticity in qualitative research. Guba and Lincoln consider validity in quantitative research as a legacy of positivism and instead propose a new ideal, "authenticity". In order to bridge the gap between the two poles of validity and authenticity, the idea of "degrees of validity" was developed during class discussions. It was hypothesized that research of an investigator (qualitative/quantitative) goes through various stages of development, in which there is a continual interaction with the community of researchers, which in the final analysis determines the validity or authenticity of the research. Most participants were quite comfortable with the idea of "degrees of validity" as neither validity in the quantitative sense nor authenticity in the qualitative sense can provide instant approval or acceptance from the community.

Chapter 6

Qualitative Methodology and Its Pitfalls in Educational Research*

Introduction

Recent research in education shows the increasing popularity of qualitative methodologies (Denzin & Lincoln, 2000, 2005; Eisner, 1997; Erickson & Gutierrez, 2002; Kivinen & Rinne, 1998; Lincoln & Guba, 2000; Macbeth, 1998; Peshkin, 2000). Lincoln and Guba (2000) consider the legitimacy of the qualitative paradigm to be not only well established but soon may even supersede that of the quantitative paradigm. Eisner and Peshkin (1990) suggest that qualitative research in education has now been institutionalized. At the same time qualitative methodology and research in education has been scrutinized critically and the need for integrating different methodologies has been emphasized (Husén, 1997; Kennedy, 1999; Mayer, 2000, 2001; Miller, Nelson & Moore, 1998; Niaz, 1997, 2008d; Saloman, 1991; Shulman, 1986). Mayer (2000) has pointed clearly to the dilemma faced by educational researchers by asking a very pertinent question: "What is the place of science in educational research?" (p. 38) and has also provided a possible/tentative answer: "scientific research can involve either quantitative or qualitative data; what characterizes research as scientific is the way that data are used to support arguments" (p. 39). An analogy from the history of science can help to illustrate this point. J.J. Thomson (1897), the celebrated British physicist, is generally credited to have "discovered" the "electron". Thomson's article in the *Philosophical Magazine* (Thomson, 1897) is an eye opener for educators as Thomson makes an extraordinary effort to present arguments (and not only data) to his peers so that they could accept a particular interpretation of the data. What, however, has generally been ignored is the fact that two other scientists (Kaufmann, 1897; Wiechert, 1897) also reported similar data at about the same time. Falconer (1987) explains why the work of Kaufmann and Wiechert was ignored, whereas Thomson was recognized as the "discoverer" of the electron:

> Kaufmann, an ether theorist, was unable to make anything of his results. Wiechert, while realizing that cathode ray particles were extremely small and universal, lacked Thomson's tendency to speculation. He could not

* Reproduced with permission from: Niaz, M. (2009c). Qualitative methodology and its pitfalls in educational research. *Quality & Quantity: International Journal of Methodology, 43*, 535–551.

> make the bold, unsubstantiated leap, to the idea that particles were constituents of atoms. Thus, while his work might have resolved the cathode ray controversy, he did not "discover" the electron.
>
> (p. 251)

In other words, data in themselves do not constitute "science", but rather it is the arguments (interpretations) scientists put forward to convince their peers that help to construct science. Interestingly, science teachers also do not understand how Thomson was able to "discover" the electron (for educational implications of this episode, cf. Niaz, 1998, 2009a). A subsequent study has gone beyond by showing how a teaching strategy based on arguments can facilitate freshman students' conceptual understanding of atomic structure (Niaz, Aguilera, Maza & Liendo, 2002).

It appears that a major problem with respect to the qualitative paradigm is its conceptualization within the Kuhnian perspective of normal and revolutionary science in which one paradigm is replaced by another (Kuhn, 1970). Guba and Lincoln (1989) refer to this Kuhnian perspective explicitly: "It is our contention that the conventional paradigm [quantitative] is undergoing a revolution in the Kuhnian sense (Kuhn, 1970), and that the constructivist paradigm [qualitative] is its logical successor" (p. 84). Lincoln (1989) is even more explicit in recognizing qualitative researchers' debt to the Kuhnian conceptualization of progress in science (pp. 60–61). The study by Miller, Nelson and Moore (1998) is a good example of how qualitative researchers still find resistance to their work and this led the authors to conclude: "We conclude that our field needs to become more reflective about practice and to develop a more deeply democratic discourse for research, one grounded in principles of academic freedom and supported by the conviction that *diversity engenders strength*" (p. 377, emphasis added). Interestingly, however, these authors ignore the fact that on the hand they espouse "diversity" and on the other subscribe to the Kuhnian thesis of paradigm "shifts" (cf. Miller, Nelson & Moore, 1998, p. 380). Diversity in this context amounts to questioning the Kuhnian thesis and has been interpreted as coexistence of competing paradigms by Shulman (1986):

> Where Kuhn erred, I believe, is in diagnosing this characteristic [controversy] of the social sciences as a developmental disability ... it is far more likely that for the social sciences and education, the coexistence of competing schools of thought is a natural and quite mature state.
>
> (p. 5)

More recently, Paul and Marfo (2001) have endorsed the use of alternative paradigms in cogent terms: "The impression created is that the much-touted paradigm shift in educational inquiry is about replacing one hegemony with another rather than giving legitimacy to the proposition that the purview of educational inquiry should be *broadened* to include alternative paradigms" (p. 538, original italics). An important pitfall about the qualitative paradigm is the misconception that qualitative research does not use quantitative data.[1] It is important to note that Guba and Lincoln in their published work have been emphatic with respect to the use of quantitative data in qualitative research and the following are two examples:

> qualitative methods are preferred, and *not* because these methods are the basis for defining the constructivist paradigm (as they are often taken to be;...). Moreover, there is nothing in this formulation that militates *against* the use of quantitative methods; the constructivist is obviously free to use such methods without prejudice when it is appropriate to do so (for example, using a questionnaire, poll, survey, or other assessment device to gather information from a broad spectrum of individuals.
>
> (Guba & Lincoln, 1989, p. 176, original italics)

> From our perspective, both qualitative and quantitative methods may be used appropriately with any research paradigm. Questions of method are secondary to questions of paradigm, which we define as the basic belief system or worldview that guides the investigator, not only in choices of method but in ontologically and epistemologically fundamental ways.
>
> (Guba & Lincoln, 1994, p. 105)

It is important to note that Guba and Lincoln endorse the use of not only quantitative data but also quantitative methods. The difference between "validity" (Maxwell, 1992) and "authenticity" (Lincoln & Guba, 2000) as understood by quantitative and qualitative researchers, respectively, has also been the subject of considerable debate (Niaz, 2007). Similarly, generalization of results obtained from qualitative studies is a controversial issue. The objective of this study is to explore the degree to which in-service teachers understand the difference between qualitative/quantitative data and methods, validity/authenticity, generalization and how these can be used to solve problems faced by the teachers in the classroom.

Rationale and Design of the Study

This study is based on 84 in-service teachers who had enrolled for the required course, Methodology of Investigation in Education, as part of their Master's degree program. Of the 84 participants, 26 were Spanish teachers, 22 English as a second language, 17 chemistry, 9 mathematics and 10 physics teachers. Age of the participants ranged from 25 to 45 years and the teaching experience ranged from 5 to 15 years (female = 55, male = 29; high school teachers = 56, university teachers = 28). Participants were divided into three groups and the same procedure was followed in all three. Almost all participants had seen a methodology course at the undergraduate level, based on texts such as Kerlinger and Lee (2002). The main objective of the course was to go beyond and provide an opportunity to in-service teachers to familiarize themselves with the controversial nature of progress in science (growth of knowledge) and its implications for research in education. Few participants had basic knowledge of the work of Popper, Kuhn, Lakatos and other philosophers of science. Some participants had a basic understanding of constructivism but it was not grounded in any particular theoretical framework.

Course Content (Reading List)

The course was based on 11 required readings and was subdivided in the following sections:

a. *Epistemology*: 1. Montero (1992); 2. Martínez (1993);
b. *Methodology*: 3. Erickson (1986); 4. Martínez (1998); 5. Maxwell (1992);
c. *Qualitative research*: 6. Brown (1994); 7. Tobin and LaMaster (1995);
d. *Integrating qualitative and quantitative research*: 8. Glasson and Lalik (1993); 9. Cortéz and Niaz (1999).
e. *Methodological and philosophical reflections*: 10. Niaz (1997); 11. Lincoln and Guba (2000).

The reading list shows how the course was designed explicitly not only to incorporate epistemological, philosophical and methodological issues, but also alternative teaching strategies. This aspect of the course facilitated exploration of alternative approaches to methodology, referred to as a "hybrid" design by Shulman (1986, p. 4).

Course Organization and Activities

On the first day of class (2 hours) all participants were provided copies of all the readings and salient features of the course were discussed. It was emphasized that the course called for active participation. As all the teachers worked in nearby schools and universities, three types of course activities were programmed:

1. Class discussions were planned on Saturdays of the 3rd, 5th and 7th week of the course (3 hours in the morning and 3 in the afternoon). Readings 1–4 were discussed in the first meeting, readings 5–8 in the second meeting and readings 9–11 in the third meeting. Participants were supposed to have studied each of the readings before the respective meetings. Each meeting started off with various questions and comments by the participants. The instructor intervened to facilitate understanding of the issues involved. (Total time devoted to class discussions = 18 hours.)
2. Class presentations were programmed during the 10th and final week of the course (Monday to Saturday, total time = 44 hours). On the first day of class, participants had formed groups of three or four and selected one of the 11 readings for their presentation. Each of the groups was assigned 2 hours (1 for presentation and the other for interventions and discussions). The presenters were supposed to moderate the discussions. The instructor

intervened when a deadlock was reached on an issue. It was expected that the participants would present the important aspects of the readings, with the objective of generating critical discussions. All groups prepared overheads based on PowerPoint presentations.

3. During class presentations in the final week teachers were encouraged to ask their questions in writing, which were then read out loud by the moderator. The presenters were given the opportunity to respond and then a general discussion followed. At the end of each session all written questions were submitted to the instructor, which provided important feedback with respect to issues, conflicts and interests of the teachers. Each participant signed her/his question. The same procedure was followed in all 11 presentations, generating a considerable amount of data (a pool of 840 questions). Besides this data, the instructor also took class notes throughout the course. This data facilitated the corroboration (triangulation of data sources) of teachers' responses to the six research assertions in this study.

Evaluation

All participants presented an Initial exam in the first session of the 10th week and a Final exam during the last session of the 10th week. Both exams were "open-book" (about 3 hours each) and participants were allowed to consult any material that they felt could be helpful. There was no particular reason for labeling these evaluations as "Initial" or "Final", except for the fact that university regulations used these labels. However, the rationale behind these evaluations (which could have been named differently) was not merely evaluation in the traditional sense. Initial exam reflected participants' experience based on 18 hours of class discussions during the first 9 weeks. Final exam reflected an additional experience of 36 hours of class presentations and discussions during the final week. Furthermore, both exams were based on the premise that oral presentations and discussions can be complemented by written responses and thus facilitate greater understanding. This study is based on participants' responses to Item 1 of the Final exam, based on six research assertions:

> Consider the following questions with respect to research in education and you are asked to respond by indicating: (a) In agreement; (b) In partial agreement; and (c) In disagreement:
>
> 1. Qualitative research as a method can help us to solve all the problems in education.
> 2. Quantitative data cannot be used and much less its statistical treatment.
> 3. It is not necessary to establish the validity of research but rather its "authenticity" or in other words interpretative validity.
> 4. It is not desirable, nor necessary to generalize from results obtained in qualitative studies.
> 5. If the phenomena are observed directly (participant observation) greater is the "authenticity" or validity of the research.

6. Formulation of hypotheses, manipulation of variables, and the quest for causal relations do not help to solve problems.

Note: Please justify your response in each case.

Multiple Data Sources

Based on different course activities, this study generated the following data sources:

a. Question–answer sessions after each of the 11 formal presentations, in which participants wrote their questions/comments (interventions), which were then discussed in class and later submitted to the instructor (a pool of 840 questions and comments was generated).
b. Initial and Final exams during the final week, separated by 36 hours of class presentations and discussions.
c. Instructor's class notes based on the following activities throughout the course: class discussions during the 3rd, 5th and 7th week of the course, 11 formal presentations during the final week, question–answer sessions after each presentation.

At this stage it is important to emphasize the role played by multiple data sources in educational research. Given the nature of paradigm wars (Gage, 1989; Howe, 1988), educational literature has suggested the need to move beyond the quantitative versus qualitative research designs and called for mixed methods research, namely: "researchers should collect multiple data using different strategies, approaches and methods in such a way that the resulting mixture or combination is likely to result in complementary strengths" (Johnson & Onwuegbuzie, 2004, p. 18). Recent research considers the mixed methods research as a new and emerging paradigm (Johnson & Christensen, 2004; Johnson & Turner, 2003; Niaz, 2008d; Onwuegbuzie & Leech, 2005; Sale & Brazil, 2004). Qualitative researchers have also generally endorsed triangulation of data sources and Guba and Lincoln (1989) consider that: "triangulation should be thought of as referring to *cross-checking specific data items* of a factual nature (number of target persons served, number of children enrolled in a school-lunch program…)" (p. 24, emphasis added). More recently, Guba and Lincoln (2005) have clarified that although qualitative and quantitative paradigms are not commensurable at the philosophical level (i.e., basic belief system or worldview), still, "within each paradigm, mixed methodologies (strategies) may make perfectly good sense" (p. 200). Interestingly, however, Shulman (1986) had advocated the need for hybrid designs much earlier: "These hybrid designs, which mix experiment with ethnography, multiple regressions with multiple case studies, process-product designs with analyses of student mediation, surveys with personal diaries, are exciting new developments in the study of teaching" (p. 4). Within this perspective it is plausible to suggest that this study has a hybrid research design. Research reported here is based on actual classroom practice and participants did not feel constrained by the research design. All the data generated (62 hours of class activities) was based on regular classroom activities. Example of a similar study is provided by Niaz (2004a).

Results

In the following sections participants' responses to the six research assertions of Item 1 of the Final exam are presented:

Qualitative Research and Problems in Education

Results reported in this section are based on participants' responses to Research Assertion 1: "Qualitative research as a method can help us to solve all the problems in education" (see Table 6.1). This was a fairly extreme and implausible proposition and it can be observed that 50% of the participants disagreed. At the other extreme, 11% of the participants agreed with this research question.

One of the participants who agreed with the research question provided the following justification:

> The nature of human beings is to interpret what surrounds them. Taking into consideration that it is natural for a human being to be interpretative, it is the intention of the researcher to interpret human educational realities by implementing interpretative procedures.

Another participant who partially agreed with the proposition, reasoned:

> Qualitative research is more inductive and leads to the comprehension of particular situations. Participant observation facilitates the understanding of social and cultural values of those participating in the design of strategies that reflect multiple realities. However, such strategies do not resolve all the problems, as at times we have to use controlled experiments and statistical treatments.

In contrast, a participant who disagreed gave the following justification: "Educational problems are complex and diverse. Methodology is helpful in the degree to which it adapts to the problem being studied. Although, qualitative research is helpful, there is no universal method that can solve all the problems." Some of the participants who disagreed referred to Montero (1992, Reading 1), who suggests that the problem precedes the method and determines the methodology to be used. Similar advice has also been offered by Saloman (1991).

These results and responses show that when participants are confronted with an extreme proposition, i.e., same methodology to solve all the problems, most of them (partially agree = 39% and disagree = 50%, total = 89%) do recognize that the method has to adapt itself to the problem situation.

Note: In order to provide reliability of results (triangulation of data sources), participants' interventions during the 11 class presentations were checked (see

Table 6.1 Participants' Responses to Research Assertions in the Study (n = 84)

	Number of Participants		
Research Assertions	*Agree*	*Partially agree*	*Disagree*
1. Qualitative research can help us to solve all the problems in education	9 (11%)	33 (39%)	42 (50%)
2. Quantitative data cannot be used and much less its statistical treatment	4 (5%)	8 (10%)	72 (86%)
3. It is not necessary to establish the validity of research but rather its "authenticity" or in other words interpretative validity	36 (43%)	17 (20%)	31 (37%)
4. It is not desirable nor necessary to generalize from results obtained in qualitative studies	36 (43%)	23 (27%)	25 (30%)
5. If the phenomena are observed directly (participant observation) greater is the "authenticity" or validity of the research	39 (46%)	26 (31%)	19 (23%)
6. Formulation of hypotheses, manipulation of variables, and the quest for causal relations do not help to solve problems	20 (24%)	20 (24%)	44 (52%)

Note
Figures in parentheses represent percentages.

pool of questions, Multiple Data Sources). It was found that participants who agreed or disagreed with this research assertion expressed similar ideas on at least two different presentations. It is important to note that Guba and Lincoln (1989) consider such "cross-checking specific data items of a factual nature" (p. 241) as an essential part of triangulation of data sources.

Quantitative Data and its Statistical Treatment

Results reported in this section are based on participants' responses to Research Assertion 2: "Quantitative data cannot be used and much less its statistical treatment" (see Table 6.1). As compared to Assertion 1, in this case the possible alternatives are stated much more clearly. In other words, if the participants accept the proposition then the use of quantitative data is excluded. Apparently, this led 86% of the participants to disagree with the proposition, and as compared to Assertion 1, the percentage of those who agreed (5%) and partially agreed (10%) also decreased considerably.

One of the participants who agreed with the research assertion stated: "In the quantitative paradigm data are collected to evaluate preconceived models, hypotheses and theories." Popularity of inductive methods fostered by some of the qualitative researchers led some participants to express concern with respect to the role played by preconceived hypotheses. Interestingly, during class discussions of the role played by philosophy of science (Readings 1 and 2), it was emphasized how observations are theory-laden (Hanson, 1958; Kuhn, 1970; Lakatos, 1970). Another participant partially agreed with the proposition and justified it in the following terms:

> Educational problems depend on the epoch, the region, the community, different ethnic groups, and hence deal with very particular issues. Nevertheless, it is instructive to compare regions and countries, and for that we need quantitative data and its statistical treatment.

One of the participants disagreed with the proposition and reasoned cogently: "If quantitative data are generally impregnated with descriptions, then descriptions (an important characteristic of qualitative research) based on the two types of research can corroborate each other." Such responses show a better appreciation of how observations are theory-laden. Another participant, who also disagreed, went beyond by recognizing that:

> According to Lakatos, growth of knowledge is facilitated by competing research programs. This means that we can use quantitative data and its statistical treatment, and at the same time have interpretations based on qualitative research ... In other words, the two paradigms can coexist in order to provide validity to our research findings.

Note: In order to provide reliability of results (triangulation of data sources) participants' interventions during the 11 class presentations were checked (see pool of questions, Multiple Data Sources). It was found that all participants who agreed or disagreed with this research assertion expressed similar ideas on at least two different presentations.

Validity of Research or Authenticity

Results reported in this section are based on participants' responses to Research Assertion 3: "It is not necessary to establish the validity of research but rather its 'authenticity' or in other words interpretative validity" (see Table 6.1). Both validity and "authenticity" was the subject of considerable debate during class discussions, especially in Readings 5 and 11. Both Maxwell (1992) and Lincoln and Guba (2000) being qualitative researchers do not agree with the ideal of validity as utilized in quantitative research. However, Maxwell differentiates between different forms of validity and makes an attempt to find common ground between qualitative and quantitative research. Lincoln and Guba (2000), on the other hand, consider validity in quantitative research as a legacy of positivism and instead propose a new ideal, namely, "authenticity". In order to bridge the gap between the two poles of validity and authenticity, the idea of "degrees of validity" was developed during class discussions. It was hypothesized that research of an investigator (qualitative/quantitative) could go through the following phases: (a) preparation of project for a dissertation, that is first revised by the advisor and approved by the research committee; (b) presentation of the dissertation, which is approved by the committee after an oral defense and corrections/changes; (c) presentation of the thesis at a congress in a specialized field; (d) publication in a journal, preferably indexed in *Current Contents* (Institute of Scientific Information, ISI); (e) impact factor of the publication as reflected in *Social Science Citation Index* (ISI, Thomson-Reuters); (f) finally, the award of a Nobel Prize or similar prizes or testimony of recognition by peers and the community at large. An important aspect of these phases is that there is a continual interaction with the community of researchers, which in the final analysis determines the validity or authenticity of the research. Furthermore, it is recommended that indicators such as impact factor be used along with qualitative criteria (Garfield & Welljams-Dorof, 1992). Most participants felt quite comfortable with the idea of "degrees of validity" as neither validity in the quantitative sense nor authenticity in the qualitative sense can provide instant approval or acceptance from the community. Interestingly, Lakatos (1974) has referred to this as how one cannot learn from experience about the truth of any scientific theory and that instant rationality comes close to being a chimera.

One of the participants who agreed with the proposition reasoned:

> Based on Mishler (1990), Campbell & Stanley (1963) and Cook & Campbell (1979), I conclude that validity cannot be determined by following procedures but rather depends on the judgments of the researchers, precisely because it is the interpretations that are significant, i.e., authenticity or interpretative validity.

Another participant who partially agreed with the proposition justified in the following terms:

> A researcher may consider his work to be authentic due to its relevance for solving certain problems ... this, however is not a guarantee that the scientific community would accept that. As we know, validity in the final analysis is determined by the community.

An example of a participant who disagreed is provided by the following:

> It is always necessary to establish not only the authenticity but also the validity of a research. All investigations can be authentic, but they need to have validity as well, which is provided by reaching a consensus among the evaluators. In my opinion, the "degree of validity" obtained by a work makes it more authentic and inspires confidence.

It is interesting to observe that those who agreed (43%) with the research question emphasized authenticity more than validity. On the other hand, those who agreed partially (20%) and disagreed (37%) recognized that validity is a relatively long-term process in which the community plays an important role.

Note: In order to provide reliability of results (triangulation of data sources) participants' interventions during the 11 class presentations were checked (see pool of questions, Multiple Data Sources). It was found that those participants who agreed or disagreed with this research assertion expressed similar ideas on at least two different presentations.

Generalizability in Qualitative Research

Results reported in this section are based on participants' responses to Research Assertion 4: "It is not desirable nor necessary to generalize from results obtained in qualitative studies" (see Table 6.1). It is generally accepted by qualitative researchers that generalizability is neither desirable nor necessary, as such studies are not designed to allow systematic generalizations to some wider population. Freidson (1975) has expressed this dilemma in cogent terms: "There is more to truth or validity than statistical representativeness" (p. 272). In spite of the difficulties of generalization in qualitative research, Maxwell (1992) has clarified that, "This is not to argue that issues of sampling, representativeness, and generalizability are unimportant in qualitative research" (p. 293). Maxwell then goes on to differentiate between: (a) internal generalizability, namely, generalizing within the community, group or institution studied to persons, events and settings that were not directly observed or interviewed; and (b) external generalizability, namely, generalizing to other communities, groups or institutions.

Table 6.1 shows that 43% of the participants in this study agreed with the research question and the following is an example of justification provided:

> Generalization in qualitative research is based on the development of a theory, which interprets not only the persons and situations under study, but also shows that the same processes in different situations can lead to different results. It is generally believed that scientific knowledge is based on objectivity, i.e., based on data and experience. On the other hand, without the mediation of the researcher these data cannot be organized into scientific knowledge. But there are also those who insist that there is a profounder knowledge that is based on intuition and speculation—a notable example of such a thinker is Bergson … This means that the researcher while attempting to present a totally objective, precise, valid and general knowledge, will never be able to do so, as his account must have some degree of subjectivism based on his traditions, myths, beliefs and sentiments.

This response goes beyond the dilemma of the degree to which results obtained in qualitative research are generalizable. In other words, reporting of data/results from both qualitative and quantitative research inevitably involves interpretation, which often leads to controversies. Philosophers of science have referred to this as the under-determination of theory by evidence (cf. Duhem–Quine thesis; Kuhn, 1970; Lakatos, 1970; Quine, 1953). Progress in science has been witness to many such controversies, in which the same data has been interpreted differently by two scientists.[2] Interestingly, Polanyi (1964), a philosopher-physical chemist, has expressed similar ideas in cogent terms:

> Our vision of reality, to which our sense of scientific beauty responds, must suggest to us the kind of questions that it should be reasonable and interesting to explore. It should recommend the kind of conceptions and empirical relations that are intrinsically plausible and which should therefore be upheld, even when some evidence seems to contradict them, and tell us also, on the other hand, what empirical connections to reject as specious, even though there is evidence for them—evidence that we may as yet be unable to account for on any other assumptions.
>
> (p. 135)

Similarly, Martin Perl (Nobel Laureate in Physics, 1995), presently working on the isolation of fractional electrical charges (quarks, which perhaps reopens the Millikan–Ehrenhaft controversy), has developed a philosophy of speculative experiments in which he is quite categorical with respect to methodological decisions:

> Choices in the design of speculative experiments [cutting-edge] usually cannot be made simply on the basis of pure reason. The experimenter usually has to base her or his decision partly on what feels right, partly on what technology they like, and partly on what aspects of the speculation they like.
>
> (Perl & Lee, 1997, p. 699)

This clearly shows how the plausibility of some conceptions (theory/generalization) can be upheld even if there is evidence to the contrary. In other words, a community of researchers may decide to interpret empirical evidence in a particular way, even if there are alternative interpretations available. This suggests that even in qualitative research generalizability is possible, provided we are willing to grant that our conceptions/theories are not entirely grounded on evidence but rather on the degree to which the community can uphold such a consensus. Campbell (1988b) has expressed this concern for objectivity, on the one hand, and the process of peer interaction, on the other, in cogent terms:

> The objectivity of physical science does not come from turning over the running of experiments to people who could not care less about the outcome, nor from having a separate staff to read the meters. It comes from a social process that can be called competitive cross-validation … and from the fact that there are many independent decision makers capable of rerunning an experiment, at least in a theoretically essential form.
>
> (p. 324)

Table 6.1 shows that 27% of the participants partially agreed with the research assertion and one of the participants alluded to the difficulty of generalizing in qualitative research due to the incommensurability thesis (Kuhn, 1970) in the following terms: "It is difficult to generalize in qualitative research as data and theories may be incommensurable, i.e., lack of communication. On the other hand, in quantitative research, generalization is more feasible as theories are commensurable."

Table 6.1 shows that 30% of the participants disagreed with the research assertion and one of the participants justified in the following terms: "Generalizability does not refer to what is repeated many times (results and findings), but rather what belongs to the person. If we have studied the person well, generalization will always manifest itself through some particular person."

This response raises an important issue that was discussed in class, especially in Reading 4 (Martínez, 1998). For example, Shakespeare's portrait of Lady Macbeth and Piaget's original studies based on his children represent findings that were not repeated many times in order to be generalized. In other words, a researcher's insight and acumen can facilitate generalizations that are later corroborated by the community.

Note: In order to provide reliability of results (triangulation of data sources), participants' interventions during the 11 class presentations were checked (see pool of questions, Multiple Data Sources). It was found that those participants who agreed or disagreed with this research assertion expressed similar ideas on at least two different presentations.

Participant Observation and Authenticity in Qualitative Research

Results reported in this section are based on participants' responses to Research Assertion 5: "If the phenomena are observed directly (participant observation) greater is the 'authenticity' or validity of the research." Table 6.1 shows that 46% of the participants agreed with the proposition and the following is an example:

> If the phenomenon is observed closely (participant observation), there is less possibility of it being refuted. How can we doubt the position of someone who is directly involved in observing the phenomenon? Is it possible that the opinion of other investigators could carry more weight than that of the participant? Definitely no.

Table 6.1 shows that 31% of the participants partially agreed with the proposition and the following are two examples:

> To emphasize observation is to give more importance to "particulars", i.e., induction. To particularize is not always the most adequate strategy, as it limits our knowledge and theoretical development.

> The researcher must be aware that there are limits to "participant observation" as it may interfere in the development of investigative processes. In any case with respect to validity, it is the community that takes care of it.

Table 6.1 shows that 23% of the participants disagreed with the proposition and following is an example of justification provided by one of the participants: "Observations are always permeated with some theory. It does not matter who makes the observations and independent of the distance between the observer and the observed." Responses to this assertion show that it elicited the highest percentage (46%) of "agree" responses. Similarly, it elicited the lowest percentage (23%) of "disagree" responses. As compared to other questions, participants found this proposition to be the most "agreeable". This is all the more interesting as during class discussions the controversy between Millikan and Ehrenhaft with respect to the oil drop experiment[3] was discussed (Holton, 1978a, 1978b; Niaz, 2000a, 2003). Ironically, Millikan and Ehrenhaft perhaps observed more drops and more closely than anybody else and still their interpretations were entirely different. Similarly, Matthews (1994a) has argued that by emphasizing observations researchers are perhaps approximating the Aristotelian ideal of empirical science.

Note: In order to provide reliability of results (triangulation of data sources) participants' interventions during the 11 class presentations were checked (see pool of questions, Multiple Data Sources). It was found that those participants who agreed or disagreed with this research assertion expressed similar ideas on at least two different presentations.

Quantitative Research and Problem Solving

Results reported in this section are based on participants' responses to Research Assertion 6: "Formulation of hypotheses, manipulation of variables and the quest for causal relations do not help to solve problems." This proposition basically asked the question as to whether quantitative research can help to solve problems in education. Table 6.1 shows that 24% of the participants agreed with the proposition and following are two examples:

> When we elaborate a hypothesis, it means that we have already defined the problem. When we determine variables and manipulate them with a certain intention, it means that we are consciously driving the research towards a certain objective. When we look for causal relations, it means the incorporation of ideas that will make the inclusion of other aspects difficult. The elements mentioned here help to operationalize the scientific method, which becomes an obstacle in problem solving.
>
> The steps mentioned in this proposition constitute the different steps of the dominant paradigm, namely, experimental research. Following this scientific method, at times problems can be solved. However, sometimes in order to construct science one has to abandon the scientific method, and this is what Millikan did in his famous oil drop experiment.

This response and many others considered experimental research to be the same as the "scientific method". Millikan, for example, definitely departed from the scientific method, but at the same time formulated presuppositions (working hypotheses), manipulated variables, looked for causal relations and did not abandon his hypotheses in the light of evidence to the contrary.

Table 6.1 shows that 24% of the participants partially agreed with the proposition and following is an example:

> When an investigator starts to observe a phenomenon, he immediately associates it with a particular theory in order to organize his work. In the determination of the elementary electrical charge, if Millikan and Ehrenhaft had come to an agreement, today we would not be investigating about "quarks"[4] as the problem could have been solved much earlier.

This response lacks the understanding that progress in science inevitably leads to controversy and hence the need for validation by the community. Furthermore, Millikan and Ehrenhaft in 1910–1915, did not have the technical means or the know-how necessary for having postulated "quarks".

Table 6.1 shows that 52% of the participants disagreed with the proposition and the following is an example:

> Many of our "universal laws" that have been accepted by the "scientific community" were the product in the first instance of reflections, doubts, anxieties with respect to observed phenomena. This was later followed by the formulation of hypotheses, presuppositions, manipulation of variables and finally the postulation of causal relations facilitated the resolution of numerous problems.

Interestingly, this response comes quite close to what philosophers of science have referred to as "idealization" in scientific progress. For example, Millikan's oil drops and Piaget's epistemic subject are epistemological constructions, namely, abstractions from the evidence of experience and are valid only when a considerable number of disturbing factors itemized in the *ceteris paribus* clauses are eliminated (cf. Kitchener, 1986, 1993; Matthews, 1994a; McMullin, 1985; Niaz, 1991a).

Note: In order to provide reliability of results (triangulation of data sources) participants' interventions during the 11 class presentations were checked (see pool of questions, Multiple Data Sources). It was found that those participants who agreed or disagreed with this research assertion expressed similar ideas on at least two different presentations.

Conclusion

This study shows that in-service teachers have considerable difficulty in understanding various critical issues with respect to educational research. Most participants had a fairly good idea of the underlying issues of the following research assertions and hence either disagreed or agreed only partially (see Table 6.1 for data): (1) Qualitative research can help us to solve all the problems in education; (2) Quantitative data cannot be used and much less its statistical treatment; and (6) Formulation of hypotheses, manipulation of variables and the quest for causal relations do not help us to solve problems. In contrast, the percentage of participants who agreed on the following research assertions was fairly high and hence problematic: (3) It is not necessary to establish the validity of research but rather its "authenticity" or in other words interpretative validity; (4) It is not desirable nor necessary to generalize from results obtained in qualitative studies; and (5) If the phenomena are observed directly (participant observation) greater is the "authenticity" or validity of research.

Most participants understood and agreed that the problem to be investigated precedes the method and determines the methodology to be used (Research Assertion 1). Participants were particularly aware of the fact that as all observations are theory-laden, it is preferable that interpretations based on both qualitative and quantitative data be allowed to compete in order to provide validity to our research findings (Research Assertion 2). The difference between validity and authenticity, the subject of Research Assertion 3, was controversial and most participants considered the need for interpreting data and hence favored authenticity. Class discussions led to the idea of "degrees of validity" as both validity in the quantitative sense and authenticity in the qualitative sense ultimately depend on critical appraisals of the community. Research Assertion 4 dealt with the controversial topic of generalizability of results obtained from qualitative studies and most participants agreed that it is not desirable to generalize. Class discussions helped to grapple with the issue and suggested an alternative: the plausibility of some conceptions (theory/generalization) can be upheld even if there is empirical evidence to the contrary. This suggests that in both qualitative and quantitative research generalizability is possible, provided we are willing to grant that our conceptions/theories are not entirely grounded in empirical evidence but rather on the degree to which the community can uphold such a consensus. Most teachers considered the use of participant observation in qualitative research as non-controversial (Research Assertion 5). Class discussions based on the observations of Millikan and Ehrenhaft in the oil drop experiment facilitated the understanding that most observations are open to diverse interpretations. Matthews (1994a) has argued that qualitative researchers by emphasizing observations are perhaps approximating the Aristotelian (rather than Galilean) ideal of empirical science:

> why it was that the supposed isochronism of the pendulum was only seen in the sixteenth century [Galileo], when thousands of people of genius and with acute powers of observation had for thousand of years been pushing children on swings, and looking at swinging lamps.
>
> (p. 111)

Finally, many teachers considered the formulation of hypotheses, manipulation of variables and the quest for causal variables (Research Assertion 6) as equivalent to the scientific method. Class discussions facilitated the understanding that such a sequence of steps does not necessarily constitute an algorithm but rather helps to reduce the complexity of a problem by controlling variables (idealization) that are related to the *ceteris paribus* clauses.

Next Chapter

In response to one of the research questions in this chapter, some teachers considered the formulation of hypotheses, manipulation of variables and the quest for causal variables as equivalent to the scientific method. Chapter 7 explored teachers' understanding and the ability to formulate hypotheses and predictions. Results obtained showed that even after some training, participating teachers had considerable difficulty in formulating and differentiating between a hypothesis and a prediction. It is concluded that although one may not subscribe to positivism as a philosophical framework (step-wise algorithms), the ability to elaborate and differentiate between these concepts provides greater comprehension and sharpens students' critical understanding of progress in science.

Chapter 7

Did Columbus Hypothesize or Predict?*

Facilitating Teachers' Understanding of Hypotheses and Predictions

Introduction

Burns and Dobson (1981, p. 93), authors of a research textbook on experimental psychology and statistics, have stated that: "Columbus had a theory that the world was round. From this theory he hypothesized that if he sailed due West he would arrive at the Indies." Lawson, Reichert, Costenson, Fedock and Litz (1989, p. 683) have pointed out that Burns and Dobson made a mistake by using the word "theory" in place of hypothesis and "hypothesis" in place of prediction. Although, some researchers may consider this confusion to be merely a question of semantics, Lawson et al. (1989) consider that it weakens the scientific structure of the research:

> In reviewing science education research, there seems to be considerable confusion between an hypothesis (a tentative explanation) and a prediction (an expected result). Hypotheses arise in response to the initial causal questions, whereas predictions generally arise deductively from the relationship between the hypothesis and the experimental procedure.
>
> (p. 683)

Although one could argue that this conceptualization comes quite close to "the scientific method" in the positivist tradition (Burbules & Linn, 1991; Niaz, 1994), still in our opinion it points out an important problem in educational research. Most philosophers of science question the existence of "the scientific method", as a series of specifiable procedures that could form an algorithm (Cartwright, 1983; Feyerabend, 1975; Giere, 1999; Kuhn, 1962, Lakatos, 1970; Polanyi, 1964). In other words, the ability to elaborate and differentiate between observations, hypotheses and predictions is important even if these terms do not constitute an algorithm. In spite of their differences most philosophers would, for example, agree that observations are theory-laden, and that theories and laws are tentative in nature (Hanson, 1958). Furthermore, philosophers of science have provided greater insight into the role played by observations, hypotheses, theories and laws in the construction of knowledge.

* Reproduced with permission from: Niaz, M. (2004b). Did Columbus *hypothesize* or *predict* that if he sailed due West, he would arrive at the Indies? *Journal of Genetic Psychology, 165*(2), 149–156.

Research in science education has shown that even science major freshman students have considerable difficulty with hypothetico-deductive reasoning (Adey & Shayer, 1994; Lawson et al., 1991; Niaz, 1996b, 1996c). In a recent study, Cortéz and Niaz (1999) have studied adolescents' (6th to 11th grade, 11–17-year-olds) understanding of *observation*, *prediction* and *hypothesis*. It was found that even the 11th-grade students (with the best performance) had considerable difficulties and obtained a mean score of 47.6% on everyday items and of 37.3% on educational context items. This shows that students have a better chance of differentiating between *predictions* and *hypotheses* on items based on everyday contexts. It is important to note that Columbus' discovery of America (referred to above) is definitely an everyday topic (starting at home and in school at a very early age) in almost all North and South American countries.

The objective of this chapter (first part) is to investigate high school and university teachers' ability to elaborate and understand the difference between *hypotheses* and *predictions*, in the everyday context of Columbus' discovery of America.

Rationale and Design of the Study

Participants

Participants were 83 in-service teachers enrolled for the required course Methodology of Investigation in Education, as part of their Master's degree program, at a major university in Venezuela. Age of the participants ranged from 25 to 45 years and their teaching experience ranged from 5 to 15 years (female = 55, male = 28; high school teachers = 56, university teachers = 27). Participants were divided into three groups and the same procedure was followed in all three. Almost all participants had seen a methodology course at the undergraduate level, based on texts such as Kerlinger (1975). The main objective of this course was to go beyond and provide an opportunity to in-service teachers to familiarize themselves with the controversial nature of progress in science (growth of knowledge) and its implications for research in education. The study by Cortéz and Niaz (1999) was one of the required readings in the course, which deals with the development of *observations, predictions and hypotheses* in adolescents within a philosophy of science perspective.

Test Item to Evaluate the Difference between a Hypothesis and a Prediction

The test item was elaborated after discussion with other researchers in the group and pilot-tested with three graduate students. All participants were asked to respond to the following test item as part of their course evaluation (Initial exam, open book):

> In contrast to some of his contemporaries, Christopher Columbus believed that the earth was round, and hence he thought that if he sailed

due West he would reach the Indies. Based on this, how would you formulate the "hypothesis' and "prediction" of Columbus' research project? Please justify your response and note that this item is based on Cortéz and Niaz (1999).

Procedure

On the first day of class (2 hours) all participants were provided copies of all the 11 readings (including Cortéz and Niaz, 1999), and salient features of the course were discussed. As all the teachers worked in nearby schools and universities, two types of course activities were programmed:

1. Class discussions were planned on Saturdays of the 4th and 7th weeks of the course (3 hours in the morning and 3 in the afternoon). Readings 1–5 were discussed in the 4th week and readings 6–11 during the 7th week (Cortéz & Niaz, 1999, was discussed in this meeting). Each meeting started with various questions and comments by the participants. The instructor intervened to facilitate understanding of the issues involved.
2. Class presentations were programmed during the 9th and final week of the course (Monday to Saturday, 8 hours daily, total time = 48 hours). On the first day of class presentations, participants presented an "open-book" Initial exam and the test item presented above (to evaluate the difference between a prediction and a hypothesis) formed part of this exam.

A detailed description of a similar course can be obtained in the literature (Niaz, 2004a).

Criteria for Grading Participants' Responses

In order to grade participants' responses on the test item of the Initial exam, the author and another researcher evaluated the responses of eight participants (selected randomly) as correct/incorrect. The researcher had experience in the epistemology of science with various publications relevant to the topic under study. Of the eight responses, the author and the researcher coincided on the evaluation of seven for hypothesis and seven for prediction. All differences were resolved by discussion. After this experience the author evaluated the responses of the remaining participants.

Results and Discussion

Table 7.1 shows participants' performance in the elaboration of *hypothesis* and *prediction*. It can be observed that *hypothesis* and *prediction* were elaborated correctly by 34.5% and 40.9% of the participants, respectively (the difference is statistically not significant, $\chi^2 = 0.41$). Only 20 (24.1%) participants had both responses correct, which perhaps indicates the degree to which there was consistency in understanding. Physics teachers had the best performance (70%) in

Table 7.1 Participants' Performance in the Elaboration of *Hypothesis* and *Prediction* (n = 83)

		Correct Elaboration of			
		Hypothesis		*Prediction*	
Participants	n	n	%	n	%
Chemistry	17	–	–	5	29.4
English	22	9	40.9	8	36.4
Mathematics	8	5	62.5	5	62.5
Physics	10	7	70.0	5	50.0
Spanish	26	8	30.8	11	42.3
Total	83	29	34.5	34	40.9

Note
Participants were grouped by the courses they taught.

elaborating a hypothesis, and mathematics teachers in prediction (62.5%). Some participants pointed out that according to Cortéz and Niaz (1999) the test item referred to everyday context, "as who would not have heard of Columbus' voyages to America". Interestingly, Cortéz and Niaz (1999) have reported that understanding of 11th-grade high school students (17-year-olds) of *observation, prediction and hypothesis* varied from 47.6% on everyday context items to 37.3% on educational context items. It appears that the understanding of both groups may have some similarity. Some of the participants, however, presented definitions of *hypothesis* and *prediction* (which could be considered as correct), but could not elaborate the correct responses in the context of Columbus' voyages and following are some of the examples:

> *Hypothesis*: An affirmation that has to be corroborated through studies and experiments/Suppositions that have not been confirmed and could be accepted provisionally/Anticipate events that could be corroborated later.
>
> *Prediction*: To assume something that could occur in the future/something that can occur as a consequence of accepting a hypothesis.

Some of the participants who elaborated correctly both the *hypothesis* and *prediction* explicitly pointed out that Columbus' research program involved hypothetico-deductive reasoning, and the following is an example:

> Perhaps through some form of hypothetico-deductive reasoning prevalent in the daily life of Columbus, he thought that if a navigator sailed due West and did not fall into a vacuum—in the past it was thought that the earth was flat, and the sea could end in a vacuum—he could arrive at the Indies and then return to his starting point, and hence confirm the hypothesis that the world is round.

Do participants confuse *hypothesis* with *prediction* (Burns & Dobson, 1981, p. 93, used the term *hypothesis* in place of *prediction*)? Interestingly, 42 participants explicitly classified a *prediction* as a *hypothesis*, and the following are three examples:

> If Columbus navigated towards the West, it is possible to arrive at the Indies ... (classified as an *hypothesis*, whereas it is a *prediction*).
>
> He could have arrived at the Indies, if he had sailed towards the West...
>
> Hypothesis: If Columbus navigated towards the West, he would arrive at the Indies [and then the response continued]: This is a deductive hypothesis, as it has its origin in a theory, namely, Columbus' belief that the world was round.

Interestingly, the last response has the same reasoning structure as that of Burns and Dobson (1981, p. 93), that is, used "theory" in place of *hypothesis* and "hypothesis" in place of *prediction*.

Of the 42 participants who classified the *prediction* as a *hypothesis*, 15 then went on to elaborate the *prediction* in almost exactly the same manner and were not cognizant of the contradiction. The following is an example: "Hypothesis: Navigating towards the West, Columbus could reach the Indies", and then went on to elaborate: "Prediction: If Columbus had navigated towards the West, he would have arrived at the Indies." Except for a small difference, the two have the same reasoning structure.

Conclusion

Results obtained in this study (first part) show that in-service teachers have considerable difficulty in the elaboration of a hypothesis and a prediction. Written responses provided evidence of the fact that most teachers (~60%) do not understand the difference between a hypothesis and a prediction. It was also observed that many teachers did provide a satisfactory description of what they considered to be a hypothesis and a prediction. However, the difficulty for the teachers consisted in operationalizing (elaborating and understanding) the difference between a hypothesis and a prediction in an everyday context (Columbus' voyages to America). Similar to Burns and Dobson (1981), about 50% of the teachers explicitly elaborated and classified a prediction as an hypothesis. In some cases the teachers elaborated the two in the same manner without being aware of the contradiction.

Finally, this study has educational implications by showing that just like students (Cortéz & Niaz, 1999), teachers also have difficulties with the elaboration and understanding of a hypothesis and a prediction. During class discussions, many teachers pointed out that they had not conceived the idea of operationalizing *hypotheses* and *predictions*, but rather used these concepts to elaborate research projects as presented in textbooks on methodology. One of the teachers was quite explicit: "After having discussed this reading I feel that my responses would have been quite similar to those of the students, as this is what we have taught them, namely, how to define a hypothesis and a prediction." Given the importance of such concepts for all research programs it is essential that appropriate teaching strategies be implemented, which is presented in the next section (part b) of this chapter.

How to Facilitate Teachers' Understanding of Hypotheses and Predictions? (Part b)

Most researchers would consider the development in adolescence of processes such as observing, classifying, formulating hypotheses, predicting, controlling variables, experimenting, interpreting data and drawing conclusions as essential for scientific reasoning. These processes constitute an important part of the Piagetian concrete and formal operational developmental stages (Inhelder & Piaget, 1958). According to Adey and Shayer (1994):

> In the population as a whole [England and Wales], fewer than 30 percent of 16-year-olds were showing the use of even early formal operations. That means that the majority of the population was leaving school using only concrete operations.
>
> (p. 31)

Johnson and Lawson (1998) have reported that only 25% of college students in the southwestern USA were classified as formal hypothetical thinkers. Similar results have been reported for Venezuelan freshman students (Vaquero et al., 1996). Similarly, Kuhn, Amsel and O'Loughlin (1988) have recognized the importance of processes such as evaluating evidence and coordinating evidence and theories as a prerequisite to scientific thinking.

With this background, Niaz (2004b, first part of this chapter) investigated high school and university teachers' ability to elaborate and understand the difference between the concepts of *hypothesis* and *prediction* using the everyday context of Columbus' discovery of America (see Burns & Dobson, 1981, presented above). Participating teachers were tested on the following test item.

Test Item

> In contrast to some of his contemporaries, Christopher Columbus believed that the earth was round, and hence he thought that if he sailed due West he would reach the Indies. On the basis of this belief, how would you formulate the hypothesis and prediction of Columbus' research project? Please justify your response.

Results obtained showed that almost 60% of the in-service teachers (n = 83) did not understand the difference between the terms *hypothesis* and *prediction* (cf. Niaz, 2004b, for details). The test item formed part of the regular evaluation and was presented before the terms *hypothesis* and *prediction* were discussed explicitly in class.

The objective of this study (part b of this chapter) was to evaluate the effect of an experimental teaching strategy that could facilitate greater understanding of the concepts, *hypothesis* and *prediction*.

Rationale and Design of the Study

This study is based on two groups of participants (in-service teachers), referred to as the Experimental Group (n=102) and Control Group I (n=94), respectively. Both groups were quite similar (age and experience) with respect to the participants in the previous study (Niaz, 2004b). The ages of the participants ranged from 23 to 47 years and the teaching experience ranged from 3 to 18 years (women=115, men=81; high school teachers=121, university teachers=75). As in the previous study all participants were enrolled in a methodology course as part of the Master's degree program at a major university in Latin America. Control Group I and Experimental Group participants received classes in two different cities, which ensured almost no contact between the two groups. Both groups followed the same program in which Cortéz and Niaz (1999) was one of the required readings. The procedure for introducing the terms *hypothesis* and *prediction* in the Control Group I was the same as in the previous study (Niaz, 2004b), and participants were evaluated on the test item presented above during the first session of the final week, that is before the terms *hypothesis* and *prediction* were discussed. Procedure for the Experimental Group is described in the next section.

Procedure for Experimental Group

On the first day of the final week the following problems were discussed in class (adapted from Cortéz & Niaz, 1999).

Problem 1

A lot of studies have been conducted to determine the possible causes of cancer. Many of the investigators believe that smoking cigarettes produces lung cancer. Consider the following phrases and classify them as Hypothesis, Observation or Prediction:

a. The majority of the people who do not smoke do not have lung cancer.
b. When people stop smoking, they will not have lung cancer.
c. Some people do not smoke and they have lung cancer.
d. Cigarette smoking produces cancer.

Problem 2

A lot of studies have been conducted to determine the influence of sports on people's health. Many of the investigators believe that people who participate in sports will always be healthy. Consider the following phrases and classify them as Hypothesis, Observation or Prediction:

a. Some people do not participate in sports and are healthy.
b. All those who participate in sports are healthy.
c. When people stop participating in sports they will get sick.
d. Some people participate in sports and get sick.

All participants had written copies of the two problems. Each problem was read out loud by one of the participants and each of the four options was discussed in detail and participants were asked to agree or disagree with a particular classification. Most participants felt quite comfortable with the discussions and agreed with the correct classifications, after providing arguments for or against a particular classification (a, b, c or d). The whole procedure took about 1 hour of class time. During the rest of the week, during class discussions (about 36 hours), participants raised interesting issues with respect to the two problems; this provided an opportunity for feedback. Many participants pointed out that they knew how to define these terms (*hypothesis*, *prediction*), but had never been asked to classify such phrases in a particular context. Participants in the Experimental Group were evaluated on the test item (presented above) in the last session of the final week. At this stage it is important to compare the test item and the two problems as part of the experimental procedure. The two problems required classification of the various statements, whereas the test item required the elaboration of a *hypothesis* and a *prediction*, which is even more difficult. This difference between the two problems and the test item is important, in order to avoid direct transfer between the two.

Results

The procedure for grading participants' responses to the test item was the same as in the previous study (Niaz, 2004b). Table 7.2 presents performance in the elaboration of the terms *hypothesis* and *prediction* for the Experimental Group and Control Groups I and II. Results for Control Group II were obtained from the previous study (Niaz, 2004b) and were included in order to provide further comparison between the three groups and hence greater validation.

Interestingly, in the Experimental Group, physics teachers had the best performance on both *hypothesis* (66.7%) and *prediction* (77.8%). Results show that overall (Total) the Experimental Group performed better than both Control Groups I and II in the elaboration of the terms *hypothesis* (49% correct) and *prediction* (63.7% correct). The difference is, however, significant only for *prediction* (see Chi square information below):

Hypothesis: Experimental and Control Group I χ^2 (1, n = 196) = 2.31 (ns)
Hypothesis: Experimental and Control Group II χ^2 (1, n = 185) = 3.15 (ns)
Prediction: Experimental and Control Group I χ^2 (1, n = 196) = 9.74 ($p < 0.01$)
Prediction: Experimental and Control Group II χ^2 (1, n = 185) = 8.64 ($p < 0.01$)

In the elaboration of *hypothesis*, Experimental Group participants performed better than Control Group I for English, mathematics, physics and Spanish

Table 7.2 Participants' Performance in the Elaboration of the Terms *Hypothesis* and *Prediction*

	Percentage correct elaboration of Hypothesis and Prediction		
	Experimental Group	*Control Group I*	*Control Group II*
Participants	n = *102*	n = *94*	n = *83*
Chemistry	n = 9	n = 10	n = 17
• Hypothesis	–	33.0%	–
• Prediction	66.7%	50.0%	29.4%
English	n = 49	n = 19	n = 22
• Hypothesis	55.1%	47.3%	40.9%
• Prediction	61.2%	47.3%	36.4%
Mathematics	n = 12	n = 26	n = 8
• Hypothesis	41.7%	38.4%	62.5%
• Prediction	75.0%	34.6%	62.5%
Physics	n = 9	n = 6	n = 10
• Hypothesis	66.7%	33.3%	70.0%
• Prediction	77.8%	66.7%	50.0%
Spanish	n = 23	n = 33	n = 26
• Hypothesis	52.1%	33.3%	30.8%
• Prediction	56.5%	33.3%	42.3%
Total	n = 102	n = 94	n = 83
• Hypothesis	49.0%	37.2%	34.5%
• Prediction	63.7%	40.4%	40.9%

Notes
1. Data for Experimental and Control Group I were obtained in this study.
2. Data for Control Group II were obtained in the previous study (Niaz, 2004b, first part, this chapter).

teachers. In the elaboration of *prediction*, Experimental Group participants performed better than Control Group I for chemistry, English, mathematics, physics and Spanish teachers. It is important to note that the percentage of teachers who responded correctly to both *hypothesis* and *prediction* increased from 24.5% in Control Group I to 38.2% in the Experimental Group.

Discussion and Conclusion

Given a particular context, in-service teachers have considerable difficulty in the elaboration of a hypothesis and a prediction. This may sound familiar to anybody who has worked on facilitating hypothetico-deductive reasoning among adolescents. Results obtained in this study show that even a simple experimental procedure can improve in-service teachers' understanding of a hypothesis and a prediction. Nevertheless, despite training, almost 50% of the teachers still had difficulty in formulating a hypothesis and 40% in formulating a prediction. Written responses provided further evidence of the fact that most teachers do not understand the difference between a hypothesis and a prediction. It was also found that many teachers provided a satisfactory description of what they considered to be a hypothesis and a prediction. However, a major difficulty for the teachers was operationalizing (elaborating and understanding) the difference between a hypothesis and a prediction. Similar to Burns and Dobson (1981), about 40% of the teachers explicitly elaborated and classified a prediction as a hypothesis. In some cases, the teachers elaborated the two terms in the same way without being aware of the contradiction. It is concluded that more sustained efforts are required to improve understanding of these important terms, in order to provide greater comprehension and a critical appraisal of progress in science.

Next Chapter

In this chapter teachers became aware of the difference between a hypothesis and a prediction and also had the opportunity to formulate these concepts in a particular context. Chapter 8 provided teachers a novel experience in which leading scientists interpreted the same experimental findings with different hypotheses. This provided an opportunity to understand "science in the making" in which formulating a hypothesis in itself is not an end, but rather an attempt to provide plausible explanations for experimental observations. In other words, rival hypotheses look for support in the scientific community and thus facilitate greater understanding leading to conceptual change.

Chapter 8

Facilitating Teachers' Understanding of Alternative Interpretations of Conceptual Change*

Introduction

Research in science education has recognized the importance of conceptual change based on different philosophical models (Chi, 1992; Clement, Brown & Zietsman, 1989; Dori & Hameiri, 2003; Dykstra, Boyle & Monarch, 1992; Gunstone, Gray & Searle, 1992; Neressian, 1989; Niaz, 1995b; Niaz & Chacón, 2003; Posner, Strike, Hewson & Gertzog, 1982; Taber, 2001; Vosniadou, 1994; White, 1993; Zoller & Tsaparlis, 1997). It is important to note that most of these models have drawn inspiration from various philosophical and historical sources (Giere, 1988, 1999; Kuhn, 1970; Lakatos, 1970; Laudan, 1977; Toulmin, 1961). Posner et al. (1982, p. 214) suggested the following necessary conditions for accommodation of cognitive structures leading to conceptual change as a rational sequence: (a) dissatisfaction with existing conceptions; (b) a new conception must be intelligible; (c) a new conception must appear initially plausible; and (d) a new concept should suggest the possibility of a fruitful research program. In contrast, recent research has also emphasized the role of extra-logical, affective, motivational, intuitive and contradictory factors in conceptual change (Cobern, 1996; Demastes, Good & Peebles, 1995; Lee et al., 2003; Niaz, Aguilera, Maza & Liendo, 2002; Pintrich, Marx & Boyle, 1993; Southerland, Johnston & Sowell, 2006; Strike & Posner, 1992).

The model proposed by Posner et al. (1982) and the revised model (Strike & Posner, 1992) draw heavily on the Kuhnian conceptualization of change in scientific development. For Kuhn (1970) scientific progress is based on the displacement of one paradigm by another and different paradigms are incommensurable, namely, core beliefs of scientists do not permit rational debate among different research programs (see Chapter 1 for details). On the contrary, the Lakatosian framework considers competing research programs as essential for progress and at the same time commensurable. Students' alternative conceptions (misconceptions) if interpreted as paradigms in the Kuhnian sense lead to situations that are not conducive to debate as Kuhn's incommensurability thesis implies that any one science can accommodate only one paradigm. A Lakatosian conceptual change teaching strategy, on the contrary, would consider students' alternative conceptions and scientific theories as competing research programs.

* Reproduced with permission from: Niaz, M. (2006). Facilitating chemistry teachers' understanding of alternative interpretations of conceptual change. *Interchange, 37*(1–2), 129–150.

Research in science education, psychology and philosophy of science has recognized similarities between the reasoning processes of students and scientists (Chinn & Brewer, 1993; Duschl & Gitomer, 1991; Karmiloff-Smith & Inhelder, 1976; Kitchener, 1986, 1987; Neressian, 1989; Piaget & Garcia, 1989; Glasersfeld, 1989). It is plausible to suggest that in the case of the Thomson–Rutherford, Bohr, Millikan–Ehrenhaft, caloric–kinetic theory controversies (all these episodes are discussed in this study), an entirely logical approach did not help the scientists to achieve consensus as they were all guided by the "hard-core" (Lakatos, 1970) of their theoretical frameworks. It is not far-fetched to suggest that conceptual change is difficult to achieve, among other reasons, precisely due to students' adherence to the "hard-core" of their epistemological beliefs. Niaz (2000c) has shown how even freshman students, after having responded correctly in one context that approximates the kinetic view of heat energy, fall back on the caloric theory of heat (hard-core of beliefs) in a different context. In the light of this framework, teaching strategies would have to be "anchored" not only in students' formal operational reasoning ability (Piaget, 1985) but also experiences that can enrich their "cognitive repertoire" and thus loosen the grip of the "hard-core" of beliefs. Providing students with alternative/conflicting views that constitute rival theories for students' thinking can facilitate such an experience.

Progress in science itself has been witness to controversies (alternative/conflicting views) and Machamer, Pera and Baltas (2000) have expressed this cogently:

> Many major steps in science, probably all dramatic changes, and most of the fundamental achievements of what we now take as the advancement or progress of scientific knowledge have been controversial and have involved some dispute or another. Scientific controversies are found throughout the history of science.
>
> (p. 3)

The same authors, however, point out that, paradoxically, "While nobody would deny that science in the making has been replete with controversies, the same people [scientists and philosophers] often depict its essence or end product as free from disputes, as the uncontroversial rational human endeavor par excellence" (Machamer et al., 2000, p. 3). This clearly shows how the role of controversy between rival/conflicting views has been ignored not only by science educators but also scientists and philosophers.

The objective of this chapter is to facilitate in-service chemistry teachers' understanding of conceptual change based on alternative philosophical interpretations (controversies/conflicting views).

Rationale and Design of the Study

This study is based on 17 in-service teachers who had enrolled in the course Investigation in the Teaching of Chemistry, as part of a Master's degree program in education at a major university in Latin America. Nine teachers worked in secondary schools and eight at university level (male = 6, female = 11; age range:

25–45 years), and their teaching experience varied from about 5 to 20 years. In the previous year all teachers had enrolled in the course Methodology of Investigation in Education, in which basic philosophical ideas of Popper, Kuhn and Lakatos were discussed in order to provide an overview of the controversial nature of progress in science (growth of knowledge) and its implications for research methodology in education. Teachers were familiar with the notion that basic ideas like the scientific method, objectivity and empirical nature of science were considered to be controversial and questionable by philosophers of science.

Course Content (Reading List)

The course was based on 17 required readings and was subdivided into the following sections:

Unit 1: *History and Philosophy of Science in the Context of the Development of Chemistry*: 1. Matthews (1994b); 2. Adúriz-Bravo et al. (2002); 3. Solbes and Traver (1996, 2001); 4. Leite (2002); 5. Niaz (1998); 6. Niaz (2000a).
Unit 2: *Students' Alternative Conceptions*: 7. Furió et al. (2002); 8. Sanger and Greenbowe (1997); 9. De Posada (1999); 10. Kousathana and Tsaparlis (2002); 11. Niaz (2000c); 12. Campanario and Otero (2000).
Unit 3: *Conceptual Change in Learning Chemistry*: 13. Niaz (1995b); 14. Niaz et al. (2002); 15. Niaz (2002); 16. García (2000); 17. Marín (1999).

The reading list shows that the course was designed explicitly to incorporate not only important areas of research in chemistry education but also a critical appraisal of the research methodology, based on controversies and conflicting views.

Course Organization and Activities

On the first day of class (2 hours) all participants were provided copies of all the readings and salient features of the course were discussed. It was emphasized that the course called for active participation. As all teachers worked in nearby schools and universities, three types of course activities were programmed:

1. Class discussions were planned on the Saturdays of the 4th, 5th and 6th week of the course (3 hours in the morning and 3 in the afternoon). Readings 1–6 were discussed in the first meeting, readings 7–12 in the second meeting and readings 13–17 in the third meeting. Teachers were supposed to have studied each of the readings before the meetings. Each meeting started with various questions and comments by the participants. The instructor intervened to facilitate understanding of the issues involved. (Total time devoted to class discussions = 18 hours.)
2. Class presentations were programmed during the 11th and final week of the course (Monday to Saturday, total time = 44 hours). On the first day of class all participants selected (by a draw) one of the 17 readings for a presentation. Each participant was assigned 90 minutes (30 minutes for the presentation

and 60 minutes for interventions and discussions). Each of the presentations was moderated by one of the participants. The instructor intervened when a deadlock was reached on an issue. It was expected that the participants would present the important aspects of the readings, with the objective of generating critical discussions. All participants prepared overheads based on PowerPoint presentations.

3. During class presentations in the final week, participants were encouraged to ask their questions in writing, which were then read out loud by the moderator. The presenters were given an opportunity to respond and then a general discussion followed. At the end of each of the sessions all written questions were submitted to the instructor, which provided important feedback with respect to issues, conflicts and interests of the participants. The same procedure was followed in all 17 presentations, generating a considerable amount of data (a pool of about 325 questions). This data facilitated the corroboration of participants' responses to the three research questions in this study.

Conceptual Change in the History of Science Discussed in Class

The following episodes from the history of science were discussed explicitly in order to demonstrate that even scientists face considerable difficulty when faced with alternative/conflicting interpretations that require conceptual change:

1. J.J. Thomson propounded the hypothesis of "compound scattering", according to which the deflection of an alpha particle by a large angle resulted from successive collisions between the alpha particle and the positive charges distributed throughout the atom. Thomson's interpretation was, of course, based on the "hard-core" of his conceptual framework, namely, the "plum-pudding" model of the atom. E. Rutherford, in contrast, propounded the hypothesis of "single scattering", according to which a deflection by a large angle resulted from a single collision between the alpha particle and the massive positive charge in the nucleus. Rutherford's interpretation, in turn, was also based on his conceptual framework, namely, the nuclear atom. The controversy between the two led to a bitter dispute. Interestingly, many textbooks consider the postulation of Rutherford's model of the nuclear atom as a *very logical consequence of the data*, and yet Thomson (acknowledged world master in the design of atomic models, cf. Heilbron & Kuhn, 1969, p. 223) did not find Rutherford's interpretation logical. This episode was discussed in Readings 5 and 14.
2. In order to explain the paradoxical stability of the Rutherford model of the atom, N. Bohr postulated the "quantum of action". Many leading physicists rejected Bohr's conceptual framework, i.e., they refused to give up their own framework and thus did not accept conceptual change. For example, Otto Stern objected, "If that nonsense is correct which Bohr has just published [Bohr, 1913], then I will give up being a physicist" (reproduced in Holton, 1986, p. 145). This episode was discussed in Readings 5 and 14.

3. Acceptance of the elementary electrical charge was preceded by a bitter controversy between R.A. Millikan and F. Ehrenhaft that lasted for many years (1910–1925). Both Millikan and Ehrenhaft obtained very similar experimental results and yet Millikan was led to postulate the elementary electrical charge (electrons) and Ehrenhaft to fractional charges (sub-electrons). Holton (1978a) has demonstrated how both Millikan and Ehrenhaft subscribed to two different conceptual frameworks, namely, atomism and anti-atomism, respectively. Resolution of the controversy was not necessarily a logical process; rather, slowly the scientific community found more merit in Millikan's interpretation. Ehrenhaft (1941), in his last published work, still argued for sub-electrons. Discussion of this episode was based on Reading 6.
4. Differentiation between heat energy and temperature has been the subject of considerable research in science education. According to Einstein and Infeld (1938/1971):

   > The most fundamental concepts in the description of heat phenomena are *temperature* and *heat*. It took an unbelievably long time in the history of science for these two to be distinguished, but once this distinction was made rapid progress resulted.
   >
   > (p. 36)

   Similarly, Brush (1976) has emphasized that "the kinetic theory could not flourish until heat as a substance [caloric theory] had been replaced by heat as atomic motion" (p. 8). Niaz (2000c) has shown that an epistemological belief in the caloric theory of heat forms part of the "hard-core" of students' framework and conceptual change requires considerable cognitive restructuring. Discussion of this episode was based on Reading 11.

Evaluation

All participants presented an Initial exam in the first session of the 11th week and a Final exam during the last session of the 11th week. Both exams were "open-book" (about 3 hours each) and participants were allowed to consult any material that they felt could be helpful. This study specifically deals with three research questions based on participants' responses to Items 2 and 3 of the Initial exam and Item 3 of the Final exam. All three research questions were based on feedback provided by the participants during different phases of the course activities.

Research Question I

This research question was based on Item 2 of the Initial exam and stated:

> According to Campanario & Otero (2000), Reading 12: "students generally ignore (metacognition) that they have wrong previous knowledge about the content of the topic under study and the reasoning processes developed in

learning science are not adequate. If a student thinks that scientific knowledge consists of facts, formulae, and data then his disposition to use cognitive resources in learning and comprehending science would be different from that of a student who has more adequate epistemological conceptions" (pp. 165–166). How would you solve this dilemma in your class?

Research Question 2

This research question was based on Item 3 of the Initial exam and stated:

> According to Marín (1999), Reading 17: "Undoubtedly, the sequence *dissatisfaction—conflict—exposition of a new idea—conceptual change*, is not all that evident in the context of the students' cognition" (p. 88, original italics). Please respond to the following questions:
>
> a. Do you agree with this critique with respect to conceptual change in learning chemistry? Explain.
> b. What arguments would you provide in favor of or against the idea of conceptual change?

Research Question 3

This research question is based on Item 3 of the Final exam and stated:

> According to Niaz (2002), Reading 15: "As a prerequisite for conceptual change it is essential that students be provided with opposing views that apparently contradict their previous thinking (alternative conceptions), and the two constitute rival theories for students' thinking" (p. 246). Please respond to the following questions:
>
> a. Do you agree with this thesis? Explain.
> b. What arguments would you provide in favor of or against this thesis?

Triangulation of Data Sources

This study is based on two main sources of data: (a) participants' written responses to the three research questions as part of the Initial and Final exams. This data helped to substantiate the research questions and consequently the educational implications; (b) participants' written questions during the 17 class presentations (a pool of approximately 325 questions, see part c of Course Organization and Activities). This data served to corroborate participants' responses to the three research questions and thus facilitated triangulation. The two sets of data represented considerable amount of arguments, controversies and critical discussions with respect to conceptual change. Participants felt free to defend a thesis or present arguments for the rebuttal of an alternative.

Results and Discussion

In this section, results obtained from participants' responses to the three research questions are presented.

Students' Epistemological Conceptions

This section is based on participants' responses to Research Question 1. Campanario and Otero (2000), Reading 12, present a critique of research with respect to the alternative conceptions of students in different content areas. They suggest that researchers must go beyond by making students cognizant of these alternative conceptions and reasoning processes (metacognition) based on an epistemological framework. In a nut-shell, if a student thinks that chemistry is about data, facts and formulae then his epistemological framework would not facilitate conceptual understanding. Hence it is important that both students and teachers be aware of alternative epistemological frameworks that can facilitate their understanding beyond that of simple diagnosis of alternative conceptions. Participants showed considerable interest during discussion of this reading and generally there was consensus with respect to the need for taking research beyond what is generally reported with respect to alternative conceptions in the literature. The following three examples illustrate participants' thinking:

> By confrontation of students' actual epistemological conceptions with more adequate conceptions. This task is facilitated by interactive argumentation, for example in a context in which a student has to formulate a hypothesis...

> To understand that science does not signify only formulae and experiments, we will have to abandon the conductivist legacy of our teachers ... One way of doing this is that while teaching a topic the following pertinent question be dealt with: how did we come to have a certain piece of knowledge. This will require a historical reconstruction of the subject and facilitate greater conceptual understanding.

> Strategy of metacognition amounts to "learning to learn", namely, the teacher must know the contradictions between the students alternative conceptions and those of the scientists ... one way of doing this is to provide students a whole series of possible responses so that they can argue and reason and ultimately have a better conceptual understanding.

Note: In order to provide reliability (triangulation of data sources) of these results, participants' interventions during the 17 class presentations (which included written questions, see pool of questions) were checked. It was found that similar ideas were expressed by 12 of the participants on at least two presentations.

More recently, Campanario (2002) has elaborated cogently with respect to the resistance offered by both scientists and students to new scientific ideas:

> The history of science can be used with a *metacognitive* dimension. This implies taking advantage of episodes of resistance to conceptual change in the history of science in order to stimulate students intellectual curiosity and make them more conscientious of their own misconceptions that result in resistance to change.
>
> (p. 1107, original italics)

Alternative Interpretations of Conceptual Change: Inconsistencies, Conflicts, Resistances and Change

Research Questions 2 and 3 provided a considerable source of conflicts, arguments and discussions for the participants. It is important to note that Research Question 2 was presented at the first session of the final week of the course, whereas Research Question 3 was presented at the last session, and between the two the participants devoted almost 40 hours to reflections, discussions, criticisms, arguments and counter-arguments. This experience enriched their understanding of the issues involved and Table 8.1 provides a profile of how participants agreed/disagreed to the propositions in Research Questions 2 and 3. Participants' responses were classified into categories depending on whether they agreed, partially agreed or disagreed on Research Questions 2 and 3 (see Table 8.2).

The results shown in Tables 8.1 and 8.2 show that Research Questions 2 and 3 posed a considerable challenge to the participants and hence the wide range of responses. For example in category (a) of Table 8.2, the nine participants showed some degree of inconsistency by agreeing to the two research questions, namely,

Table 8.1 Profile of Participants' Responses to Research Questions 2 and 3 (n = 17)

Participant	*Research Question 2*	*Research Question 3*
1	Agreed	Disagreed
2	Partially agreed	Agreed
3	Agreed	Agreed
4	Agreed	Agreed
5	Agreed	Agreed
6	Partially agreed	Agreed
7	Agreed	Agreed
8	Agreed	Agreed
9	Partially agreed	Agreed
10	Agreed	Agreed
11	Disagreed	Disagreed
12	Partially agreed	Partially agreed
13	Agreed	Agreed
14	Agreed	Agreed
15	Agreed	Agreed
16	Agreed	Partially agreed
17	Disagreed	Agreed

Table 8.2 Classification of Participants' Responses in Categories Depending on Whether They Agreed, Partially Agreed or Disagreed on Research Questions 2 and 3

Category	*Research Question 2*	*Research Question 3*	*No. of Participants*
a	Agreed	Agreed	9
b	Partially agreed	Agreed	3
c	Disagreed	Agreed	1
d	Agreed	Partially agreed	1
e	Partially agreed	Partially agreed	1
f	Agreed	Disagreed	1
g	Disagreed	Disagreed	1

Note
Inconsistent categories: a, b and d (see text for details).

agreement with Research Question 2 implies that cognitive conflicts do not necessarily lead to conceptual change, whereas an agreement with Research Question 3 implies that under certain circumstances (presence of rival theories) cognitive conflicts can facilitate conceptual change. Similarly, categories (b) and (d) also demonstrated some degree of inconsistency. It is plausible to suggest that the experience provided by the course facilitated a majority of the participants (13 out of 17, categories a, b and d) to reconsider and thus change (deepen) their understanding of conceptual change. Nevertheless, it is important to note even those participants who used consistent strategies (for example, category (f)) also raised important issues with respect to the implementation of conceptual change.

Of the nine participants in category (a), at least six explicitly experienced an inconsistency and recognized extra-logical, novel and original attempts to resolve contradictions leading to conceptual change.

Note: In order to provide reliability of these results, interventions of these nine participants during the 17 class presentations (that included written questions, see pool of questions) were checked. It was found that these participants expressed similar ideas on at least two (or more) different presentations. The following are two examples of participants' responses in category (a), i.e., agreement with both Research Questions 2 and 3.

First Example Category (a)

Response to Research Question 2—agreed in the following terms: "What constitutes a conflict for the teacher may not be perceived as such by the student so as to produce a cognitive conflict leading to equilibration (assimilation—accommodation) of cognitive structures in the Piagetian sense."

Response to Research Question 3—agreed in the following terms:

> It appears that within the Lakatosian framework, just like the scientist a student can live with two theories simultaneously—later as the student acquires

> some expertise the conflict can perhaps be resolved. Furthermore, this coincides with Mischel's (1971) requirement that the student must engender his own conflicts in order to cope with a problem, and this precisely motivates cognitive development.

This example clearly shows a deeper understanding of conceptual change. In Research Question 2 the participant had some reservations with respect to how a conflict would be perceived and later resolved by the student. Research Question 3 provides an alternative, namely, the student may not perceive the conflict at the same time and context as the teacher, but leaves open the possibility (somewhat intuitively) of working simultaneously with two different "theories" which at some later stage may facilitate conceptual understanding. This shows that the teacher must be aware that students can respond to conflicting situations in various forms. Chinn and Brewer (1993) provide a series of possible strategies used by students.

Second Example Category (a)

Response to Research Question 2—agreed in the following terms: "A supposed cognitive conflict strategy meticulously designed by the teacher may not be construed as such by the student ... A student after all is not a scientist."

Response to Research Question 3—agreed in the following terms:

> The ideas of Kuhn and Lakatos permeate this question ... A teacher can "manipulate" so as to present the scientifically accepted idea as a rival to students' alternative conceptions ... in this particular context conceptual change does not follow immediately as a logical solution, but rather the student experiences different frameworks at the same time leading to a conflict and perhaps later the construction of a solution, and nor does cognitive theory explain all the situations ... this is quite similar to what happens to scientists before the scientific community decides in favor of one or the other theory. An important aspect of this thesis is that students take their time to construct the new framework ... Nevertheless, teachers' epistemological outlook is crucial in such situations.

This response clearly recognizes the role of different philosophical interpretations, i.e., straightforward introduction of cognitive conflict or the possibility of considering two different rival ideas, within an extra-logical situation, that is difficult to explain cognitively. Furthermore, in order for the latter to be effective, the teachers' epistemological views are crucial.

Example Category (b)

Response to Research Question 2—partially agreed in the following terms: "It represents a logical sequence and is feasible—but perhaps it would deviate from what is expected as conceptual change is very complex. It is not easy to produce changes in the "hard-core" epistemological beliefs of students."

Response to Research Question 3—agreed in the following terms: "Exposing students to cognitive conflicts that lead them to entertain rival ideas can facilitate conceptual change as it stimulates comprehension by facilitating a dynamic atmosphere based on participation and the strengthening of a critical attitude."

This response recognizes that it is difficult to change "hard-core" (Lakatos, 1970) epistemological beliefs. Nevertheless, a dynamic and critical atmosphere in the classroom can facilitate conceptual change.

Example Category (e)

Response to Research Question 2—partially agreed in the following terms:

> it is not easy for the students to be aware of their alternative conceptions and hence they may not feel dissatisfied. On the contrary, students generally are quite convinced that what they have learned is correct ... this makes conceptual change difficult.

Response to Research Question 3—partially agreed in the following terms: "my concern is that the opposing view presented to the student must be the correct scientific idea, so that he/she can decide for himself whether it coincides or is contrary to the previous thinking." (This participant had expressed similar ideas on two different presentations, see pool of questions.)

This response alludes to the fact that students' alternative conceptions offer resistance to change (especially if they form part of the "hard-core", cf. Chinn & Brewer, 1993) and that somehow students must be aware of what is the "correct scientific idea". This concern is valid as some social constructivists (cf. Tobin & LaMaster, 1995) have suggested that what the students have learned (including alternative conceptions) is much more important than what they should have learned, namely, the "correct scientific idea". It is important to note that this group of teachers had discussed various social constructivist perspectives, including that of Tobin and LaMaster (1995), in their Methodology course in the previous year.

Example Category (f)

Response to Research Question 2—agreed in the following terms: "the sequence outlined in the research question does not seem to be feasible ... in general students are satisfied with their ideas, whether right or wrong."

Response to Research Question 3—disagreed in the following terms:

> there cannot be just one way of producing conceptual change ... It is possible that a student may achieve conceptual change by his own investigations, explorations, confrontations and imagination, without the need to introduce an external conflict ... in science there is no absolute truth and consequently there cannot be rules, methods, algorithms, steps for introducing conceptual change.

(This participant had expressed similar ideas on two different presentations, see pool of questions.)

This response raises issues that are crucial for a better understanding of conceptual change. The crux of the issue is that if history and philosophy of science have shown us that the scientific method we teach to our students is a caricature of what scientists actually do (cf. Fuller, 2000, p. 212), then how can we elaborate and accept a certain sequence, steps, algorithms or rules for introducing conceptual change in the classroom? Based on Readings 5 (Niaz, 1998) and 6 (Niaz, 2000a), participants were particularly aware that most textbooks schematize progress in chemistry as based on the scientific method, namely, a scientist observes and experiments, and as a result enunciates a theory that is later confirmed repeatedly and finally the law is enunciated.

Conclusions

According to Marín (1999), Reading 17 (p. 88), the sequence (dissatisfaction—conflict—exposition of a new idea—conceptual change) approximates quite closely to the original scheme suggested by Posner et al. (1982) and, furthermore, there is no evidence to show that this may always lead to conceptual change. This in our opinion is a plausible critique. Marín (1999), in his critique, however, goes beyond by suggesting that students' reasoning abilities (formal operational cognitive structure, Piaget, 1985) are perhaps more important for conceptual change (cf. Marín, Benarroch, Gómez, 2000 for further elaboration on this point). This contrasts sharply with some historical controversies in the history of science. For example, Thomson did not find Rutherford's interpretation of alpha particle experiments as logical, despite chemistry textbooks' claim to the contrary (cf. Niaz, 1998, Reading 5). Similarly, Bohr's postulation of the "quantum of action" was not accepted as a logical alternative by many renowned physicists (cf. Holton, 1993). The Millikan–Ehrenhaft controversy provides an eloquent case of two scientists who obtained very similar experimental data and still "stuck" to their presuppositions (hard-core of beliefs) to provide two entirely different interpretations. Interestingly, at one stage in the controversy, Millikan stated:

> That these same ions have one sort of charge when captured by a big drop and another sort when captured by a little drop is obviously *absurd* ... Such an assumption is not only too *grotesque* for serious consideration but is directly contradicted by my experiments.
>
> (Millikan, 1916, p. 617, emphasis added)

A student may wonder as to whether experimental data can be considered as "absurd" and "grotesque" (Niaz, 2003). A major finding of this study is that most teachers went through an experience based on historical controversies that involved: inconsistencies, conflicts, contradictions and finally some degree of conceptual change. A few of the participants although resisted any change, but still raised important issues with respect to conceptual change. Based on participants' responses to the three research questions, this study has important educational implications:

1. Students' prior and alternative epistemological conceptions play an important role in learning chemistry;
2. Instead of emphasizing formulae and experiments (as end products of scientific endeavor) it is preferable to make students think as to "how did we come to have a certain piece of knowledge?";
3. Students generally are quite convinced that what they have learned is correct and this makes conceptual change difficult;
4. Need for going beyond the simple diagnosis of alternative conceptions, leading to "learning to learn" strategies (metacognition);
5. What constitutes a cognitive conflict for a teacher may not be perceived as such by the student and hence the difficulties associated with the introduction of conceptual change;
6. Similar to a scientist, a student can "live" with two rival theories (ideas/concepts) simultaneously and, as the student acquires expertise or enriches his cognitive repertoire, the conflict can perhaps be resolved;
7. Under certain circumstances resolution of a conflict may not follow a logical pattern of reasoning but rather a slow process (based on motivational, intuitive and affective factors) in which the "hard-core" of beliefs slowly "crumbles";
8. In contrast to some social constructivists, it is important that students be aware as to what is the expected correct scientific idea—apparently this would help them to plan their own cognitive strategies;
9. In science there is no absolute truth, nor a "scientific method", and consequently there cannot be rules, methods, algorithms or predetermined steps for introducing conceptual change; and
10. Teachers' epistemological outlook is crucial in order to facilitate conceptual change.

Next Chapter

This chapter provided teachers with an opportunity to consider alternative hypotheses for the same experimental findings, in order to achieve conceptual change. Chapter 9 goes beyond by providing more detailed examples from "science in the making", which is far more complex than the traditional scientific method found in textbooks. Teachers became aware of the various strategies scientists use in their research programs, such as conjugation of speculation and reason as an important element in looking for an answer to a particular question, adoption of the methodology of idealization (simplifying assumptions) in order to solve complex problems, science does not develop by appealing to objectivity in an absolute sense and opening new "windows" for facilitating conceptual change.

Chapter 9

Progressive Transitions in Teachers' Understanding of Nature of Science Based on Historical Controversies*

Introduction

Recent research in science education has emphasized the importance of history and philosophy of science (HPS) (Burbules & Linn, 1991; Campanario, 2002; Hodson, 1988b; Justi & Gilbert, 1999; Matthews, 1994a, 2004; Monk & Osborne, 1997; Niaz, 1994). *Project 2061* (AAAS, 1989, 1993) and the *National Science Education Standards* (NRC, 1996) have also recognized the importance of HPS. Within the HPS perspective, nature of science (NOS) is considered to be one of the most important topics of research and the subject of considerable controversy (Abd-El-Khalick & Lederman, 2000b; Alters, 1997; de Berg, 2003; Eflin, Glennan & Reisch, 1999; McComas, Almazroa & Clough, 1998; Niaz, 2001a; Smith, Lederman, Bell, McComas & Clough, 1997; Scharmann & Smith, 2001; Smith & Scharmann, 1999; Solomon, Scott & Duveen, 1996). Given the complexity of the nature of science even for philosophers of science, some of the controversy has centered on the issue of what needs to be included and the level of complexity. According to Smith and Scharmann (1999), nature of science can help students and teachers to distinguish between things that are more scientific from those that are less scientific. Based on this premise, and in contrast to Niaz (2001a), these authors do not recommend the following aspects of NOS, namely, "scientific progress does not depend on empirical data alone and is characterized by competition among rival theories", as this would not help to distinguish science from non-science (Scharmann & Smith, 2001). More recently, Lederman et al. (2002) have suggested that we need to go beyond and "focus on individual classroom interventions aimed at enhancing learners' NOS views, rather than on mass assessments aimed at describing or evaluating students' beliefs" (pp. 497–498). Similarly, Clough (2006) has proposed that "NOS instruction would likely be more effective if teachers deliberately scaffolded classroom experiences and students' developing NOS understanding back and forth along the continuum" (p. 463). The present study is designed to meet such requirements. Inclusion and the discussion of NOS in the classroom is important as it is often associated with myths, such as universality of the scientific method, science is procedural rather than creative, scientists are particularly objective and

* Reproduced with permission from: Niaz, M. (2009b). Progressive transitions in chemistry teachers' understanding of nature of science based on historical controversies. *Science & Education, 18*, 43–65.

that experiment is the sole arbiter of scientific knowledge (cf. McComas, 1996, for details). Some readers may consider the scientific method to be associated with scientific inquiry rather than with NOS. It is important to point out that research in science education has generally considered the scientific method as part of NOS. McComas et al. (1998) have presented a consensus view of NOS based on eight international science Standards Documents in the following terms: "There is no one way to do science (therefore, there is no universal step-by-step scientific method)" (p. 513). In developing their questionnaire about NOS (Views of Nature of Science, VNOS), Lederman et al. (2002) have explicitly referred to the scientific method as a myth:

> The myth of the scientific method is regularly manifested in the belief that there is a recipe like stepwise procedure that all scientists follow when they do science. This notion was explicitly debunked: There is no single scientific method that would guarantee the development of infallible knowledge.
> (p. 501)

Again, in a Delphi study conducted by Osborne et al. (2003) the scientific method was considered to be an important part of nature of science. This is all the more important as this study was based on the responses given by science educators, historians, philosophers, sociologists and science communicators. Given these antecedents, this study (among others) has paid particular attention to the facilitation of participating teachers' understanding of the following NOS aspects: scientific method, objectivity and empirical nature of chemistry. Lederman et al. (2002, table 1, p. 506) have also emphasized these NOS aspects of their participants.

Despite the controversy a certain degree of consensus has been achieved within the science education community and NOS can be characterized by, among others, the following aspects:

1. Scientific knowledge relies heavily, but not entirely, on observation, experimental evidence, rational arguments and skepticism;
2. Observations are theory-laden;
3. Science is tentative/fallible;
4. There is no one way to do science and hence no universal step-by-step scientific method can be followed;
5. Laws and theories serve different roles in science and hence theories do not become laws even with additional evidence;
6. Scientific progress is characterized by competition among rival theories;
7. Different scientists can interpret the same experimental data in more than one way;
8. Development of scientific theories at times is based on inconsistent foundations;
9. Scientists require accurate record keeping, peer review and replicability;
10. Scientists are creative and often resort to imagination and speculation;
11. Scientific ideas are affected by their social and historical milieu.

A review of the literature shows that most teachers in many parts of the world lack an adequate understanding of some or all of the different NOS aspects outlined above (Akerson et al., 2006; Bell et al., 2001; Blanco & Niaz, 1997; Clough, 2006; Lederman, 1992; Mellado et al., 2006; Pomeroy, 1993). This should be no surprise to anyone who has analyzed science curricula and textbooks, which are heavily slanted toward an entirely empirical and positivist epistemology. Similarly, a considerable amount of work has also been done to teach NOS in the classroom (Abd-El-Khalick & Akerson, 2004; Bianchini & Colburn, 2000; Hosson & Kaminski, 2007; Irwin, 2000; Khishfe & Lederman, 2006; Lin & Chen, 2002; Oulton, Dillon & Grace, 2004). Interestingly, many science teachers would perhaps argue that NOS has no place in science courses or textbooks. Marquit (1978) has exposed this fallacy in eloquent terms: "textbooks in physics invariably have a philosophical content whether or not the authors are conscious of it,... In this way students are taught possibly highly controversial philosophical viewpoints with no indication that there are alternative outlooks" (p. 784). Teacher-training courses in NOS thus constitute a high priority for science education. Blanco and Niaz (1997) have reported that both students and teachers hold very similar beliefs about NOS. This finding is of special interest for the present study as both were based on teachers with a similar socio-economic and educational background.

Schwab (1962, 1974) differentiates between the methodological (empirical data) and interpretative (heuristic principle) components of scientific knowledge. In other words it is not verified knowledge (accumulation of empirical data) but rather the heuristic principles that help us to structure inquiry. Schwab's idea of a heuristic principle (construction of the mind) comes quite close to what modern philosophers of science have referred to as presuppositions (Holton, 1978a, 1978b), hard-core (Lakatos, 1970) or guiding assumptions (Laudan et al., 1988). Based on this perspective, Niaz (2001a) has emphasized the role of heuristic principles in understanding the nature of science. As an illustration, let us consider Thomson's (1897) determination of the mass-to-charge (*m/e*) ratio of cathode rays which helped him to postulate the existence of a universal particle, the electron. Interestingly, Thomson was neither the first to do so nor the only experimental physicist. Kaufmann (1897) and Wiechert (1897) also determined the *m/e* ratio of cathode rays in the same year as Thomson and their values agreed with each other. If we tell students that "science is empirical", we shall be denying an important aspect of the nature of science, namely what made Thomson's work different from that of Kaufmann and Wiechert. Falconer (1987) has explained the difference cogently:

> Kaufmann, an ether theorist, was unable to make anything of his results. Wiechert, while realizing that cathode ray particles were extremely small and universal, lacked *Thomson's tendency to speculation*. He could not make the bold, unsubstantiated leap, to the idea that particles were constituents of atoms. Thus, while his work might have resolved the cathode ray controversy, he did not "discover the electron".
>
> (p. 251, emphasis added)

The rationale behind the empirical determination of the *m/e* ratio was provided by the heuristic principle, namely Thomson hypothesized that if the ratio varied in different gases, cathode rays could be considered as ions, and alternatively if the ratio remained constant, cathode rays could be considered as universal charged particles. Surprisingly, most textbooks go to considerable length to describe Thomson's experimental determination of the *m/e* ratio and still ignore the heuristic principle (cf. Niaz 1998, 2009a, for details).

Various studies based on different topics of the general chemistry curriculum have shown that textbooks generally ignore the heuristic principles (construction of the mind/presuppositions) that vary from one research program to another, and the following are some of the examples: atomic structure (Niaz, 1998), determination of the elementary electric charge (Niaz, 2000a), kinetic theory (Niaz, 2000b), covalent bond (Niaz, 2001b), laws of definite and multiple proportions (Niaz, 2001c), periodic table (Brito, Rodríguez & Niaz, 2005). Besides the heuristic principles, Schwab (1962) has drawn attention to how textbooks resort to a "unmitigated *rhetoric of conclusions* in which current and temporary constructions of scientific knowledge are conveyed as empirical, literal, and irrevocable truths" (p. 24, original italics). In other words, textbooks not only ignore the heuristic principles but also emphasize a vision of science based on empirical facts that lead to scientific "truths" and ultimately represents a "rhetoric of conclusions" devoid of all reasoning, debate and controversy.

The objective of this study is to facilitate progressive transitions in chemistry teachers' understanding of NOS in the context of historical controversies. Selected controversies (although not discussed in the textbooks) referred to episodes that form part of the chemistry curriculum. The importance of teaching NOS within the context of the science curriculum has been recognized in the literature (Irwin, 2000; Niaz et al., 2002).

Rationale and Design of the Study

This study is based on 17 in-service teachers who had enrolled in the course Investigation in the Teaching of Chemistry, as part of a Master's degree program in education at a major university in Latin America. Author of this study was the course instructor. Nine teachers worked in secondary schools and eight at the university freshman level (male = 6, female = 11; age range: 25–45 years), and their teaching experience varied from about 5 to 20 years. This group of teachers had various aspects in common and formed a coherent group by sharing similar concerns, difficulties and student populations. Almost all teachers were graduates of nearby universities and had very similar educational training, which did not include any exposure to new trends in science education or philosophy of science. For example, before starting their Master's degree program, none of them had read anything by Popper, Kuhn or Lakatos. Both the secondary and university general chemistry teachers taught more or less the same topics (university level being higher) to science major students. In the previous year all teachers had enrolled in the following courses: (a) Philosophy of Education, (b) Methodology of Investigation in Education, (c) Theories of Learning, (d) Instructional Design, (e) Instructional

Strategies, and (f) Evaluation. The course Methodology of Investigation in Education was also given by the author of this study, in which basic philosophical ideas of Popper, Kuhn and Lakatos were discussed in order to provide an overview of the controversial nature of progress in science (growth of knowledge) and its implications for research methodology in education. Details of this course based on a similar group of teachers are available in the literature (cf. Niaz, 2004a). Based on this experience, teachers were familiar with the notion that basic ideas like the scientific method, objectivity and empirical nature of science were considered to be controversial and questionable by philosophers of science (Phillips, 1990, is a good example of this critique).

Course Content (Reading List)

The course was based on 17 required readings and was subdivided into the following sections:

Unit 1: *History and Philosophy of Science in the Context of the Development of Chemistry*: 1. Matthews (1994b); 2. Adúriz-Bravo et al. (2002); 3. Solbes and Traver (1996, 2001); 4. Leite (2002); 5. Niaz (1998); 6. Niaz (2000a).
Unit 2: *Students' Alternative Conceptions*: 7. Furió et al. (2002); 8. Sanger and Greenbowe (1997); 9. De Posada (1999); 10. Kousathana and Tsaparlis (2002); 11. Niaz (2000c); 12. Campanario and Otero (2000).
Unit 3: *Conceptual Change in Learning Chemistry*: 13. Niaz (1995b); 14. Niaz et al. (2002); 15. Niaz (2002); 16. García (2000); 17. Marín (1999).

The reading list shows that the course was designed explicitly not only to incorporate important areas of research in chemistry education but also a critical appraisal of the research methodology. Content and the issues discussed in the following readings were directly relevant to this study: 1, 2, 3, 4, 5, 6, 11, 13 and 14.

Course Organization and Activities

On the first day of class (2 hours) all participants were provided copies of all the readings and salient features of the course were discussed. It was emphasized that the course called for active participation. As all teachers worked in nearby schools and universities, the following course activities were programmed:

a. Class discussions were planned during the 4th, 5th and 6th week of the course (3 hours in the morning and 3 in the afternoon). Readings 1–6 were discussed in the first meeting, readings 7–12 in the second meeting and readings 13–17 in the third meeting. Teachers were supposed to have studied each of the readings before the meetings. Each meeting started with various questions and comments by the participants. The instructor intervened to facilitate understanding of the issues involved. (Total time devoted to class discussions = 18 hours.)
b. Class presentations were programmed during the 11th and final week of the course (Monday to Saturday, total time = 44 hours). On the first day of class

all participants selected (by a draw) one of the 17 readings for a presentation. Each participant was assigned 90 minutes (30 minutes for the presentation and 60 minutes for interventions and discussions). Each of the presentations was moderated by one of the participants. The instructor intervened when a deadlock was reached on an issue. It was expected that the participants would present the important aspects of the readings, with the objective of generating critical discussions. All participants prepared PowerPoint presentations.

c. During class presentations in the final week participants were encouraged to ask their questions in writing, which were then read out loud by the moderator. The presenters were given the opportunity to respond and then a general discussion followed. At the end of each session all written questions were submitted to the instructor, which provided important feedback with respect to issues, conflicts and interests of the participants. There were 17 presentations and 16 participants who asked the questions in each presentation in writing. Some participants asked more than one question and in some cases also submitted comments. Each participant signed her/his question. The same procedure was followed in all 17 presentations, generating a considerable amount of data (a pool of about 325 questions and comments). This data facilitated the corroboration of participants' responses to the three research questions in this study.

Evaluation

All participants presented an Initial exam in the first session of the 11th week and a final exam during the last session of the 11th week. There was no particular reason for labeling these evaluations as "Initial" or "Final", except for the fact that university regulations used these labels. However, the rationale behind these evaluations in this study (which could have been named differently) was not merely evaluation in the traditional sense. Both exams were "open-book" (about 3 hours each) and participants were allowed to consult any material that they felt could be helpful. This provided the participants the opportunity to reflect, think and elaborate strategies on issues closely related to classroom practice. The Initial exam reflected participants' experience based on 18 hours of class discussions during 10 weeks. The Final exam reflected an additional experience of 36 hours of class presentations and discussions during the final week. Furthermore, both exams were based on the premise that oral presentations and discussions can be complemented by written responses and thus facilitate greater understanding. This study is based on participants' responses to Item 1 of the Initial exam and Items 1 and 2 of the Final exam. In the following three sections results obtained from participants' responses to these three items are presented.

Research Questions

1. What do the teachers think with respect to myths in nature of science and chemistry education? (Based on Item 1, Initial exam.)

Item 1, Initial exam:
According to McComas et al. (1998), cited in Reading 2 (pp. 466–467), there are various myths associated with the "nature of science".

a. Do you share the thesis that there are myths with respect to the nature of science? Explain. (Research Question 1a.)
b. What other myth would you add besides those mentioned by McComas et al. (1998)? (Research Question 1b.)
c. Just as there are myths with respect to the nature of science, do you think there are myths with respect to chemistry education? (Research Question 1c.)

2. Do the teachers consider chemistry textbooks to present science as an illustration of the scientific method and what changes are necessary to improve the textbooks? (Based on Item 1, Final exam.)
Item 1, Final exam:
The scientific method is generally schematized as "Observation, experimentation, enunciation of laws and theories, confirmation of the enunciated laws and theories" (Reading 3, Solbes & Traver, 1996, p. 106).

a. Based on your experience as a teacher, do you think many of the chemistry textbooks represent science in this manner? Can you illustrate with an example? (Research Question 2a.)
b. Do you think that this is a good way to represent chemistry? (Research Question 2b.)
c. What changes would you suggest in order to improve the presentation of chemistry in textbooks and the classroom? (Research Question 2c.)

3. How do teachers understand and interpret the research methodologies of Thomson, Rutherford, Bohr, Millikan and Ehrenhaft, within the context of a philosophy of speculative experiments? (Based on Item 2, Final exam.)
Item 2, Final exam:
Martin Perl, Nobel Laureate in Physics 1995, in his search for the fundamental particle (quark), has elaborated a philosophy of speculative experiments:

> Choices in the design of speculative experiments usually cannot be made simply on the basis of pure reason. The experimenter usually has to base her or his decision partly on what feels right, partly on what technology they like, and partly on what aspects of the speculations they like.
>
> (Perl & Lee, 1997, p. 699)

Given the methodologies of Thomson, Rutherford, Bohr (Readings 5 and 14), Millikan and Ehrenhaft (Reading 6), in your opinion, what are the implications of this statement for teaching chemistry?

It is important to note that the three research questions are closely related to the three exam questions (Item 1, Initial exam; Items 1 and 2, Final exam), which are in turn a consequence of the course content and activities. In order to establish the validity of the exam questions, a research colleague was provided the course materials and the exam questions and asked to provide feedback. Based on this feedback exam questions were revised.

Multiple Data Sources

Based on the different course activities this study generated the following data sources:

a. Question–answer sessions after each of the 17 formal presentations, in which participants wrote their questions/comments (interventions), which were then discussed in class and later submitted to the instructor (a pool of 325 questions and comments was generated).
b. Initial and final written exams during the final week, separated by 36 hours of class presentations and discussions (total time = 8 hours).
c. Instructor's class notes. These were based on the following activities throughout the course: class discussions during the 4th, 5th and 6th week of the course (total time = 18 hours), 17 formal presentations during the final week, question–answer sessions after presentation.

At this stage it is important to emphasize the role played by multiple data sources in educational research. (*Note*: Teachers in this study participated in 62 hours of various class activities.) Given the nature of the paradigm wars (Gage, 1989; Howe, 1988), educational literature has suggested the need to move beyond the quantitative versus qualitative research designs and called for mixed methods research, namely, "researchers should collect multiple data using different strategies, approaches and methods in such a way that the resulting mixture or combination is likely to result in complementary strengths" (Johnson & Onwuegbuzie, 2004, p. 18). Qualitative researchers have generally endorsed triangulation of data sources and Guba and Lincoln (1989) consider that "triangulation should be thought of as referring to *cross-checking specific data items* of a factual nature (number of target persons served, number of children enrolled in a school-lunch program…)" (p. 241, emphasis added). More recently, Guba and Lincoln (2005) have clarified that although qualitative and quantitative paradigms are not commensurable at the philosophical level (i.e., basic belief system or worldview), still within each paradigm mixed methods make perfectly good sense. Interestingly, however, Shulman (1986) had advocated the need for hybrid research designs much earlier: "These hybrid designs, which mix experiment with ethnography, multiple regressions with multiple case studies, process-product designs with analyses of student mediation, surveys with personal diaries, are exciting new developments in the study of teaching" (p. 4). Within this perspective it is plausible to suggest that this study has a hybrid research design. Research reported here is based on actual classroom practice and participants did not feel constrained by the research design. All the data generated (based on 62 hours of class activities) was based on regular classroom activities.

Historical Controversies Discussed in Class

The following controversies from the chemistry curriculum were discussed explicitly:

a. Thomson's experiments were conducted to clarify the controversy with respect to the nature of cathode rays, that is, charged particle or waves in the ether (Readings 5 and 14).
b. Explanation of the large-angle deflections of alpha particles was controversial. Thomson put forward the hypothesis of compound scattering (multitudes of small scatterings), whereas Rutherford propounded the hypothesis of single scattering based on a single encounter (Readings 5 and 14).
c. Paradoxical stability of the Rutherford model of the atom, which led to Bohr's controversial thesis of incorporating Planck's "quantum of action" in the classical electrodynamics of Maxwell (Readings 5 and 14).
d. Millikan–Ehrenhaft controversy with respect to the determination of the elementary electrical charge (Reading 6).

These controversies refer to the atomic models of Thomson, Rutherford and Bohr and the elementary electrical charge (Millikan). All these topics formed part of the chemistry curriculum at the secondary and university freshman level. Readings 5 (Niaz, 1998) and 6 (Niaz, 2000a, 2000b, 2000c) deal explicitly with textbook analyses of these topics. Participants were quite familiar with most of these textbooks and the following have been used extensively as Spanish translations: Ander and Sonnessa (1981), Burns (1996), Brady and Humiston (1996), Brady and Holum (1981), Brown and LeMay (1988), Chang (1999), Ebbing (1997), Hein (1990), Mahan and Myers (1990), Masterton et al. (1985), Sienko and Plane (1971), Whitten et al. (1998). Readings 5 and 6 provide considerable historical details and specific criteria to evaluate textbooks. Reading 14 (Niaz et al., 2002), on the other hand, provided an example of a teaching strategy based on historical controversies in order to facilitate conceptual change. During classroom discussions, teachers were surprised to find as to how textbooks (that they had used for many years) could ignore important issues. The readings and the discussions thus provided considerable feedback to the participants to prepare guidelines for including these issues in the classroom. The main objective of the readings and classroom discussions was to make participants aware of alternative philosophical frameworks, besides those present in the textbooks (cf. Marquit, 1978)

Participants' Responses to Research Question 1: Myths Related to Nature of Science and Chemistry Education

Research Question 1: What do teachers think with respect to myths in nature of science and chemistry education?

It is important to note that this group of in-service teachers had their first exposure to the controversial nature of various aspects of nature of science, during the course Methodology of Investigation in Education in the previous year. However, these teachers were not aware that the difference between fact and fiction had become so blurred that now we have myths with respect to our understanding of NOS.

Participants' responses to Research Question 1a (Do you share the thesis that there are myths with respects to the nature of science?)

Fifteen participants agreed with the thesis that there are myths with respect to the nature of science (two participants gave ambiguous responses). All 15 responses to Research Question 1a were quite similar and the following are three of the examples:

> Majority of the textbooks view science from a positivist perspective, in which valid knowledge is based on the observables, the objective facts and all that can be measured. On the other hand, cognitive aspects and the constructive nature of knowledge are ignored.

> Myths can also be considered as "critical paradigms" in the Kuhnian sense ... for example, universality and rigidity of the scientific method, absolute validity of scientific knowledge are myths that respond to the changes and needs of the times.

> In a certain sense the "university" itself has perpetuated these myths that represent the needs of past epochs. On the other hand, the textbooks also emphasize such myths. The formation of teachers also helps to perpetuate these myths to a fair degree that leads to the preparation of students who are experts in solving algorithmic problems, analyzing empirical data and reception learning.

Participants' responses to Research Question 1b (What other myth would you add besides those mentioned by McComas et al. 1998?)

Thirteen participants had ideas with respect to what they considered to be other myths, besides those mentioned by McComas et al. (1998). It is possible that some of these myths may have already been reported in the literature. Nevertheless, the following examples bear testimony to the degree to which this group of teachers had followed a "progressive transition" with respect to their original thinking.

> It is often suggested that philosophical changes in the history of ideas is a consequence of advances in science. Actually, it appears that it is the philosophical changes that have led us to devote our attention to research in science [This idea is interesting if we consider that in the last 100 years many philosophical schools—positivism, Popper, Kuhn, Lakatos, Giere—have presented penetrating analyses of the same scientific episodes and still their conclusions are different].

> Reproduction of limited intellectual horizons [Perhaps only a genius can aspire to have stimulating intellectual experience].

> In view of the universality and rigidity of the scientific method, one could believe that "science does not change". For some it may signify that if science changes, *it does not exist* [original italics].

> Scientific work is a domain reserved for a few—besides majority of these is men, leaving the women in second place. Also at times one can see an "individualist perspective" in which an isolated genius is "illuminated" and thus the importance of working in groups is ignored.

Participants' responses to Research Question 1c (Just as there are myths with respect to the nature of science, do you think there are myths with respect to chemistry education?)

All (n = 17) participants took a keen interest in responding to the possibility of myths in chemistry education and the following are some examples:

> Students are made to think that they can never attain the status of a famous scientist or a genius, such as Thomson, Rutherford, Bohr and Millikan.

> Chemistry is considered to be an experimental science in which laboratory work leads to the production of knowledge with no reference to controversies and debates that help to construct scientific knowledge [This response clearly constitutes a progressive transition with respect to participants' prior belief that chemistry is primarily an empirical science].

> Learning is associated with memorization of formulae to solve algorithmic problems.

> Students believe that chemistry teachers have all the knowledge with respect to all chemistry topics and hence must have answers to all their questions. This is perhaps due to the fact that all scientific knowledge is considered to be elaborated and a finished product, and hence all their questions must have an answer.

> No effort is made to differentiate between the idealized scientific law and the observations—as a consequence students tend to memorize the laws.

Note: In order to provide reliability of results reported in this section (Research Questions 1a, 1b and 1c), participants' interventions during the question–answer sessions in the 17 class presentations were checked (which included written questions; see Multiple Data Sources in the previous section for the pool of 325 questions and comments). It was found that all participants expressed similar ideas on at least two different presentations. The objective of checking was to compare participants' responses on an issue on two different opportunities. First, their response on the written exam question and then their question/comment on a related issue during the question–answer sessions in the 17 class presentations. In the case of some uncertainty or ambiguity, instructors' class notes were also consulted. If a participant's response could not be substantiated through the different data sources, it was not included as representative of the group and we looked for an alternative. A similar procedure was followed in analyzing all exam questions. This procedure is based on the theoretical framework presented in the section on Multiple Data Sources.

Participants' Responses to Research Question 2: Scientific Method and How It Is Represented in Chemistry Textbooks

Research Question 2: Do the teachers consider chemistry textbooks to present science as an illustration of the scientific method and what changes are necessary to improve the textbooks?

At this stage it is important to note that participants' previous familiarity with the universality and rigidity of the scientific method was in a context that was not of immediate concern to them, namely, NOS as studied by philosophers of science. This research question provided an opportunity to reflect on the implications of the scientific method for chemistry textbooks and classroom practice.

Participants' responses to Research Question 2a (Based on your experience as a teacher, do you think many of the chemistry textbooks represent science in this manner?)

Fifteen teachers agreed that chemistry textbooks presented science as an illustration of the scientific method (two participants provided ambiguous responses) and the following are some examples of the responses:

> Majority of the textbooks present material based on a sequence of ordered steps that constitute the scientific method—this provides a vision of science as static and rigid.

> Textbooks present chemistry as "rhetoric of conclusions" in which the scientist has a well laid out plan with respect to what he is going to discover. The scientist observes, experiments and as a result enunciates a theory which is later confirmed repeatedly and finally the law is enunciated [for "rhetoric of conclusion", see Schwab, 1962].

> ...some textbook authors postulate the scientific method not as an alternative but rather as obligatory for the scientist...

> ...textbooks present Millikan as a "god" who knew everything and hence determined the charge of the electron—in a certain sense a pseudo-history.

> Based on the scientific method most textbooks present Bohr's postulates as a scientific theory—a teacher could wonder as to how Bohr confirmed his "postulates..." [cf. Lakatos, 1970, who considers Bohr's postulates as part of his "negative heuristic" which is based on presuppositions and generally not open to experimental refutation. Also see Chapter 12 for further elaboration].

Participants' responses to Research Question 2b (Do you think that this is a good way to represent chemistry?)

Fifteen teachers agreed that this is not a good way to teach chemistry (two participants provided ambiguous responses) and the following are some examples of the responses:

> A teacher must have not only content knowledge but also awareness of the methodological issues, history of science and its development, and the interaction between science, technology and society.

> The importance of observation and experiments is emphasized, whereas the role of construction of knowledge based on suppositions and hypotheses is ignored.

> ... science is tentative and it is the scientific community that plays the role of the arbiter ... science is not just discovery but rather construction.

Participants' responses to Research Question 2c (What changes would you suggest in order to improve the presentation of chemistry in textbooks and the classroom?)

Thirteen teachers agreed that the presentation of chemistry in the textbooks and the classroom needs to be improved and the following are some examples of the responses:

> The image of chemistry as presented in textbooks ("abstraction" and "perfection") needs to be changed ... Chemistry is not a series of laws and theories that cannot be restructured ... the difficulties faced by the scientists must be included.

> Workshops based on history and philosophy of science for chemistry teachers could help to go beyond teaching algorithms and facilitate conceptual understanding ... textbook authors need to be more receptive to the idea that scientific knowledge is tentative and based on alternative interpretations.

> Chemistry needs to be "freed" of myths and history and philosophy of science could help. It needs to be emphasized that there is no one scientific method, but rather diverse methods and processes—textbooks cannot continue to be a list of questionnaires and algorithmic problems and answers.

> The best way to teach chemistry is to narrate its history and philosophy in the context of its development—this may help to "demolish" the myth of the scientific method.

> Dramatization of historical episodes could help to generate discussions that may arouse students' enthusiasm and interest to follow a scientific career [cf. Stinner & Teichman, 2003].

Note: In order to provide reliability of results reported in this section (Research Questions 2a, 2b, 2c), participants' interventions during the question–answer sessions in the 17 class presentations were checked (that included written questions; see section on Multiple Data Sources for pool of 325 questions/comments). It was found that all participants expressed similar ideas on at least two different presentations. In the case of some uncertainty or ambiguity, instructor's class notes were consulted (cf. section on Multiple Data Sources). This procedure is based on the theoretical framework presented in the section on Multiple Data Sources.

Research Question 3: Philosophy of Speculative Experiments and Chemistry Education

Research Question 3: How do teachers understand and interpret the research methodologies of Thomson, Rutherford, Bohr, Millikan and Ehrenhaft, within the context of a philosophy of speculative experiments?

It is important to note that Martin Perl and colleagues have worked on a Millikan-style methodology in order to isolate quarks (cf. Rodríguez & Niaz, 2004b, for a comparison between Millikan's research methodology and Perl's philosophy of speculative experiments). The rationale behind using this episode from the history of science was to present an experience from a leading scientist working on cutting-edge experimental work and how a scientist goes about coping with difficulties. Thirteen participants found this item interesting and challenging, and although most presented positive implications, there were four who suggested negative implications. The following are some examples of positive implications for teaching chemistry:

> According to Lakatos, theories can "live" together for some time and after a period of arguments and confrontation the scientific community decides in favor of one or the other. Similarly, it is probable that Martin Perl considers the conjugation of speculation and reason as an important element in looking for an answer to a particular question. In the Millikan–Ehrenhaft controversy, Millikan based on the "negative heuristic" of his research program decided to discard some of the data. This was perhaps a recognition, that besides reason, speculation and intuition also played an important part ... A similar process occurred in the case of the atomic theories [Thomson, Rutherford, Bohr] ... This shows that everything cannot be solved by logic, and it is necessary to look for other alternatives provided they are consistent and well justified ... Far from confusing the students, these episodes can arouse their curiosity and hence interest in science.

> ...in the work of Thomson, Rutherford, Bohr, Millikan and Ehrenhaft besides logic, speculations played an important part ... this reconstruction based on the history of science demonstrates that scientists adopt the methodology of idealization (simplifying assumptions) in order to solve complex problems ... it is plausible to hypothesize that students adopt similar

> strategies in order to achieve conceptual understanding [for idealization, cf. McMullin, 1985; Niaz, 1999a]

> ...statement by Perl helps to "humanize" chemistry ... it opens a new window with respect to scientific knowledge ... discussion of such issues in the classroom can facilitate conceptual change towards constructivist views ... it will also require innovative teaching strategies.... [This clearly shows that inclusion of appropriate materials in teacher training courses can help to foster innovative teaching strategies—a major goal of this book.]

> The picture that emerges from these episodes shows that controversy and speculation played an important part in the construction of knowledge ... This requires the preparation of critical persons who can defend their positions ... In this regard the teacher is responsible for not inhibiting students' creativity.

> ...how many scientific advances have not been presented just because the author could not substantiate his claims based on rigorous reasoning and perhaps also the fear that the scientific community may not accept ... dissemination of the work of Millikan, Ehrenhaft and Perl among teachers ... could contribute to facilitate scientific progress.

> Millikan did not manifest in public the speculative part of his research ... Perl, however has affirmed publicly that at times he speculates ... Perl's affirmation manifests what Millikan in some sense tried to "conceal", namely, science does not develop by appealing to objectivity in an absolute sense and that science does not have an explanation for everything and hence the need for research. Acceptance of the fact that science does not have an absolute truth and nor an immediate explanation for everything, would change students' conception of science and chemistry in particular. This will show chemistry to be a science in constant progress and that what is true today may be false tomorrow and may even help to originate a new truth—sequences of heuristic principles [cf. Burbules & Linn, 1991].

Four of the participants expressed concern with respect to the inclusion of issues such as "speculation" (Perl & Lee, 1997) and pointed out negative implications for teaching chemistry and the following is an example:

> Martin Perl's affirmations may confuse teachers. In the case of Thomson, Rutherford, Bohr, Millikan and Ehrenhaft one can understand that as investigators they could base their decisions on what they felt to be correct and even speculate in order to justify their theories. One could understand this as part of the Lakatosian "negative heuristic". In the classroom, if the teacher does not understand completely the circumstances that led the scientist to adopt a particular methodology, this may produce conflicts and even an erroneous conception of science.

Note: In order to provide reliability of results reported in this section (Research Question 3), participants' interventions during the question–answer sessions in the 17 class presentations were checked (which included written questions; see section on Multiple Data Sources for pool of 325 questions/comments). In the case of uncertainty or ambiguity, instructor's class notes were also consulted. It was found that for those participants who suggested positive implications, similar ideas were expressed on at least two different presentations. For those who suggested negative implications similar ideas were expressed twice by two participants.

Progressive Transitions in Chemistry Teachers' Understanding of Nature of Science

Table 9.1 is based on participants' selected responses to the three research questions, with respect to scientific method, objectivity and the empirical nature of chemistry. Both secondary and university freshman teachers had very similar beliefs about these three concepts.

Scientific Method

Progressive transitions (see Table 9.1):

- Universality of the scientific method could signify that science does not change (Research Question 1b);
- Scientific method not as an alternative but obligatory (Research Question 2a);
- There is no one scientific method but rather diverse methods (Research Question 2c).

In Research Question 1b, the participant simply alludes to the universality of the scientific method, whereas in Research Question 2c (Final exam) there seems to be an inference based on the experience gained and diverse methods are recommended. Participants' responses to the three research questions (1b, 2a and 2c) provide a good indicator as to how their understanding changed by recognizing that there cannot be a universal scientific method but rather diverse methods. Lederman et al. (2002) have also reported similar changes in NOS views of experts and novices (see Table 2, p. 514).

Objectivity

Progressive transitions (see Table 9.1):

- Valid knowledge is based on the observables and objective facts (Research Question 1a);
- How did Bohr confirm the "postulates" of his scientific theory? (Research Question 2a);

Table 9.1 Progressive Transitions in Chemistry Teachers' Understanding of Nature of Science

Scientific method	*Objectivity*	*Empirical Nature of Chemistry*
In view of the universality of the scientific method, one could believe that: "Science does not change." For some it may signify that if science changes, *it does not exist* (emphasis in orig.) (Research Question 1b)	Majority of the textbooks view science from a positivist perspective, in which valid knowledge is based on the observables, the objective facts and all that can be measured (Research Question 1a)	Chemistry is considered to be an experimental science in which laboratory work leads to the production of knowledge with no reference to controversies and debates that help to construct scientific knowledge (Research Question 1c)
. . . some textbook authors postulate the scientific method not as an alternative but rather as obligatory for the scientist . . . (Research Question 2a)	Based on the scientific method most textbooks present Bohr's postulates as a scientific theory—a teacher could wonder as to how Bohr confirmed his "postulates" (Research Question 2a)	No effort is made to differentiate between the idealized scientific law and the observations—as a consequence students tend to memorize the laws (Research Question 1c)
Chemistry needs to be "freed" of myths and history and philosophy of science could help. It needs to be emphasized that there is no one scientific method, but rather diverse methods and processes—textbooks cannot continue to be a list of questionnaires and algorithmic problems and answers (Research Question 2c)	Workshops based on history and philosophy of science for chemistry teachers could help to go beyond teaching algorithms and facilitate conceptual understanding . . . textbook authors need to be more receptive to the idea that scientific knowledge is tentative and based on alternative interpretations (Research Question 2c)	The importance of observation and experiments is emphasized, whereas the role of construction of knowledge based on suppositions and hypotheses is ignored (Research Question 2b)

	Millikan did not manifest in public the speculative part of his research . . . Perl, however has affirmed publicly that at times he speculates . . . Perl's affirmation manifests what Millikan in some sense tried to 'conceal', namely, science does not develop by appealing to objectivity in an absolute sense and that science does not have an explanation for everything and hence the need for research (Research Question 3)	. . . how many scientific advances have not been presented just because the author could not substantiate his claims based on rigorous reasoning and perhaps also the fear that the scientific community may not accept . . . dissemination of the work of Millikan, Ehrenhaft and Perl among teachers . . . could contribute to facilitate scientific progress (Research Question 3)

Notes

1. This table illustrates progressive transitions in participants' understanding of nature of science with respect to scientific method, objectivity and empirical nature of chemistry. Each column presents the responses of the same participant throughout the course. Information in the table represents the progressive understanding of three different participants on three aspects of NOS.
2. Examples of participants' responses are taken from the two written exams (see text for more details).
3. Some of the responses have been shortened to avoid lengthy sequences.

- Scientific knowledge is not absolutely objective, hence the role of alternative interpretations and the tentative nature of science (Research Question 2c);
- If Millikan had been absolutely objective he should have published all his data. Perl acknowledges that based on his presuppositions a scientist can speculate (Research Question 3).

This sequence of progressive transitions is thought-provoking indeed. Of special interest is the question: How did Bohr confirm the "postulates" of his scientific theory? Did Bohr have the necessary empirical evidence to substantiate his "postulates"? Even a cursory glance at the historical record would reveal that Bohr's "postulates" were primarily speculative and based on his presuppositions. Bohr's famous "postulates" are given prominence in almost all chemistry textbooks. Nevertheless, textbooks generally do not discuss as to how Bohr came up with these postulates. Lakatos (1970) has explained cogently that the postulates formed part of the "negative heuristic" of Bohr's research program, which cannot be refuted by empirical evidence. The degree to which Bohr did not have empirical evidence for his postulates is evident from a letter written by Rutherford to Bohr (dated March 20, 1913):

> ...the mixture of Planck's ideas with the old mechanics makes it very difficult to form a physical idea of what is the basis of it all ... How does the electron decide what frequency it is going to vibrate at when it passes from one stationary state to another?
>
> (reproduced in Holton, 1993, p. 80)

Again, it is important to note that Rutherford was generally quite sympathetic (in contrast to many other physicists) to Bohr's model of the atom. At this stage the crucial question is, how and why did this participant ask this question in response to Research Question 2a? If we go back and read the question it will become evident that the question in itself does not mention Bohr or his postulates, but rather the scientific method in general. It is not far-fetched to suggest that this participant experienced some degree of conceptual understanding (a progressive transition) and for that reason used Bohr's example to illustrate that scientific theories do not necessarily originate with experimental observations. Interestingly, we checked participants' interventions in the question–answer sessions and found that four other participants expressed similar views.

The sequence of progressive transitions continues (see Table 9.1) and a reference is made to the methodologies of Millikan and Perl in the context of "objectivity in an absolute sense", in response to Research Question 3. The research question itself does not mention objectivity, but rather refers to the research methodologies of Thomson, Rutherford, Bohr, Millikan, Ehrenhaft and Perl. Of course, a careful reading of these historical episodes (Readings 5, 6 and 14) would show that: (a) Thomson was not convinced and did not accept that Rutherford's model of the atom followed logically from the alpha particle experiments; (b) Millikan, in his controversy with Ehrenhaft with respect to the oil drop experiment, considered some of Ehrenhaft's interpretations as "absurd" and "grotesque"; and (c) Perl in his experimental search for quarks has presented a philosophy of speculative experiments. This is the context in which this participant understood Research Question 3, and it does facilitate greater understanding with respect to the scientific endeavor and hence constitutes a progressive transition. Once again we checked participants' interventions during the question–answer sessions and found that four other participants expressed similar views. Interestingly, more recently Perl (2005), in referring to the Millikan–Ehrenhaft controversy, has pointed out,

> I agree with your conclusion that Millikan believed that the only basic charge was that on the electron and he sorted his data to get the best measurement of that charge. I don't ... [know] ... what he would have done if there were say three different particles with different values of the electric charge.

Empirical Nature of Chemistry

Progressive transitions (see Table 9.1):

- Acceptance of the role of debate and controversy in the handling of experimental data (Research Question 1c);
- Differentiation between the experimental observations and the idealized scientific laws (Research Question 1c);
- Role of suppositions and hypotheses in the construction of knowledge (Research Question 2b);
- Dilemma faced by a scientist when his suppositions are not substantiated by the experimental data (Research Question 3).

Of this sequence, perhaps the most significant is: differentiation between the experimental observations and the idealized scientific laws. The role of idealizations in the progress of science was discussed in Readings 1 (Matthews, 1994b) and 13 (Niaz, 1995b), in the context of various topics, such as Galileo's law of free fall, Newton's laws and gas laws. The basic idea that emerged from these discussions was that scientific laws being epistemological constructions do not describe the behavior of actual bodies. Nevertheless, this was considered to be a difficult topic. In spite of this, it is interesting to note that this participant invoked the idea of idealization in order to understand the relationship between empirical data and the theoretical formulations. This in our opinion constitutes fairly good evidence for a progressive transition. Once again, we checked participants' interventions during the question–answer sessions and found that two other participants expressed similar views.

It is important to note that participants in their responses referred to various aspects of NOS. However, we found the transitions with respect to scientific method, objectivity and the empirical nature of chemistry to be more representative of participants' understanding. Interestingly, Lederman et al. (2002, table 1, p. 506), in their comparison of the views of experts and novices, also draw attention to these aspects of NOS. The consistency between the three research questions and participants' progressive transitions provide evidence for the fact that the teaching approach was actually implemented as described.

Conclusion and Educational Implications

This study shows that given the opportunity to reflect, discuss and participate in a series of course activities based on various controversial episodes directly related to the chemistry curriculum, teachers' understanding of NOS can be enhanced. At the beginning of the course teachers were simply aware that ideas like the scientific method, objectivity and empirical nature of science were considered to be controversial by philosophers of science. As a next step (progressive transition) this study provided the opportunity to understand that there are myths associated with the nature of science (first research question). Participants suggested other myths besides those discussed in class, namely, limited intellectual horizon of the students (primarily due to the rigidity of the scientific method), science is a domain reserved for geniuses and men, learning is associated with memorization of formulae to solve algorithmic problems (cf. Pickering, 1990), and lack of a differentiation between idealized scientific laws and observations (cf. Niaz, 1999a). Idealization in science, namely, scientific laws being epistemological constructions do not describe the behavior of actual bodies, is considered to be "one of the major stumbling blocks to meaningful learning of science" (Matthews, 1994a, p. 211).

The second research question facilitated teachers' understanding of the scientific method within the context of chemistry textbooks. Almost all teachers agreed that chemistry textbooks presented science as an illustration of the scientific method in which Millikan is presented as a "god", there is lack of an understanding that Bohr's postulates represented the "negative heuristic" (Lakatos, 1970) of his theory, and postulation of the scientific method not as an alternative but rather as obligatory for the scientist. Teachers also suggested that presentation of chemistry in textbooks and the classroom could be improved by introducing history and philosophy of science, recognition of the role of suppositions and hypotheses in the construction of knowledge and that it is the scientific community that plays the role of the arbiter (peer review) and not the scientific method.

The third research question was the most challenging and facilitated an understanding of a philosophy of speculative experiments based on Perl and Lee (1997). It is important to note that this "philosophy" was published in the *American Journal of Physics*, which is primarily oriented toward the teaching of physics. Most teachers suggested positive implications, such as conjugation of speculation and reason as an important element in looking for an answer to a particular question; adoption of the methodology of idealization (simplifying assumptions) in order to solve complex problems; opening of a new "window" that can facilitate conceptual change toward constructivist views; science does not develop by appealing to objectivity in an absolute sense; science does not have an explanation for everything, hence the need for research. It is important to note that four teachers expressed reservations with respect to the inclusion of "philosophy of speculation" in classroom discussions as it may lead to an erroneous conception of science.

It is plausible to suggest that interactions among participants and teacher-participants in this study facilitated the following progressive transitions in teachers' understanding of nature of science:

1. Problematic nature of the scientific method, objectivity and the empirical basis of science (see Table 9.1);
2. Myths associated with respect to the nature of science and teaching chemistry (see Research Question 1);
3. Understanding of the scientific method within the context of chemistry textbooks and not just as a concern of philosophers of science (see Research Question 2);
4. The role of speculation and controversy in the construction of knowledge based on episodes from the chemistry curriculum (see Research Question 3);
5. How did Bohr confirm his postulates? This represents a novel way of conceptualizing Bohr's model of the atom, which goes beyond the treatment presented in most textbooks (see Research Question 2);
6. Science does not develop by appealing to objectivity in an absolute sense, as creativity and presuppositions also play a crucial role (see Table 9.1);
7. Differentiation between the idealized scientific law and the observations is crucial for understanding the complexity of science (see Research Question 1 and Table 9.1).

These issues have educational implications and are important for deepening teachers' understanding of nature of science. As compared to previous research, this study provides an explicit teaching strategy for introducing different aspects of NOS as part of the regular classroom activities. At this stage it is important to recall Brush's (1978) advice to chemistry teachers:

> Of course, as soon as you start to look at how chemical theories developed and how they were related to experiments, you discover that the conventional wisdom about the empirical nature of chemistry is wrong. The history of chemistry cannot be used to indoctrinate students in Baconian methods. (p. 290)

Indeed, chemistry textbooks and curriculum in most parts of the world provide "indoctrination" with respect to the scientific method, objectivity and the empirical nature of chemistry. In this context, this study provides guidelines for providing an alternative approach to understand chemistry. More recently, de Berg (2003, p. 417) has drawn attention to these aspects of teaching chemistry in cogent terms:

> Controversy often raises important philosophical issues which are not easily resolved and such controversy is particularly potent during those historical

> episodes where the nature of science itself undergoes change ... To teach chemistry in a way that acknowledges the importance of including such factors in an approach towards scientific literacy demands a reassessment and rationalization of traditional learning styles and a purposeful attempt *to incorporate these historical and philosophical factors, not as adjuncts to the course, but as important ingredients of the course.*
>
> (emphasis added)

Finally, a word of caution is necessary as the relationship between different topics of the chemistry curriculum and HPS is complex. Given the complexity of understanding NOS even for researchers in science education, it is plausible to suggest that participants in this study may not have understood the nature of science in all its complexity. Furthermore, it is essential to understand that the level of complexity at which NOS can be introduced would vary from the secondary to the freshman university level (cf. suggestions by Smith & Scharmann, 1999). However, it is important that such courses could motivate teachers to question the "conventional wisdom about the empirical nature of chemistry" and pursue further studies in nature of science within an HPS perspective.

Next Chapter

This chapter provided teachers a critical appraisal of the research methodologies of various well-known scientists, such as Thomson, Rutherford, Bohr and Millikan. Of particular interest was Martin Perl's philosophy of speculative experiments. The rationale behind using these methodologies was to present experiences from leading scientists working on cutting-edge experimental work and then show how a scientist goes about coping with difficulties. Chapter 10 went beyond by challenging teachers to deal with the horns of a trilemma. On the one hand, teachers were aware that, in order to make new discoveries, scientists need to be creative and innovative, and yet school science is generally dogmatic and authoritarian, very much in the sense visualized by Kuhn's "normal science". This clearly showed the difficulties involved in introducing NOS in the science curriculum and textbooks. Some teachers explicitly stated that Kuhn's normal science manifests itself in the science curriculum and textbooks through the scientific method. Given teachers' criticism of dogmatic and authoritarian ways of teaching science, the concern with respect to the scientific method is quite understandable.

Chapter 10

What "Ideas-About-Science" Should be Taught in School Science?*

Introduction

Research in science education has recognized not only the importance of history and philosophy of science but also the inclusion of nature of science (NOS) and "ideas-about-science" in the science curriculum, both at the secondary and university level (Matthews, 2004). Various science education reform documents have also emphasized the need for understanding NOS (AAAS, 1993; McComas & Olson, 1998; Millar & Osborne, 1998; NRC, 1996). Despite some degree of consensus as to what is the nature of science and how it can be introduced in the classroom, NOS continues to be a controversial topic. In an attempt to understand What is Science?, the American Physical Society has drafted a policy statement, which has been endorsed by the American Association of Physics Teachers in the following terms:

> Science is the systematic enterprise of gathering knowledge about the world and organizing and condensing that knowledge into testable laws and theories. The success and credibility of science is anchored in the willingness of scientists to ... abandon or modify accepted conclusions when confronted with more complete or reliable experimental evidence.
>
> (AAPT, 1999, p. 659)

For most practical purposes this definition is acceptable. Nevertheless, a closer look reveals the problematic nature of "What is science"? and hence the difficulties associated with what "ideas-about-science" should be taught in school science. For example, the emphasis on "gathering knowledge", i.e., experimental data and "complete or reliable experimental evidence" with no reference to controversy, explanation of data by rival theories, or the tentative nature of scientific knowledge shows the complexity of the issues involved (cf. Phillips, 2005a, 2005b for details). No wonder one critic considers that by endorsing this statement the members of the Executive Board of the AAPT had "classified themselves as non-scientists" (Auerbach, 2000, p. 305). Furthermore, on the basis of a history and philosophy of science (HPS) perspective, this critic points out that the word

* Reproduced with permission from: Niaz, M. (2008b). What "ideas-about-science" should be taught in school science? A chemistry teachers' perspective. *Instructional Science, 36*, 233–249.

"complete" could be construed by Kuhnians (Kuhn, 1970) as "monotheistic Popperianism" (Auerbach, 2000, p. 305), namely, scientific knowledge is moving toward an increasing absolute truth.

Although the idea of testing and hypothesizing is most germane to physical sciences like chemistry and physics, its presentation in the classroom is devoid of controversy, namely, the debatable aspects. In contrast, evolution in biology (e.g., the writings of Stephen J. Gould) provides opportunities for debate. No wonder Dobson, a physicist, has asked a very pertinent question: is physics debatable? "Can the history of physics provide such [evolution in biology] relevance? Does physics provide nice meaty controversies that might tempt the adolescent to think?" (Dobson, 2000, p. 1). Niaz and Rodríguez (2002) have argued that the inclusion of appropriate historical episodes can make physics in the classroom debatable.

In a recent Delphi-style study, Osborne, Collins, Ratcliffe, Millar and Duschl (2003) consulted 23 experts engaged in the study of science and its communication, based on the following groups: leading scientists; historians, philosophers and sociologists of science; science educators; science teachers; and public understanding of science. These experts recommended that the following "ideas-about-science" could be taught in school science: (1) scientific method and critical testing; (2) creativity; (3) historical development of scientific knowledge; (4) science and questioning; (5) diversity of scientific thinking; (6) analysis and interpretation of data; (7) science and certainty; (8) hypothesis and prediction; (9) cooperation and collaboration in the development of scientific knowledge. Interestingly, the experts assigned the highest priority to the teaching of "scientific method and critical testing", with reasons such as: (a) "core process on which the whole edifice of science is built"; (b) "central thrust of scientific research"; and (c) "careful experimentation is used to test hypotheses".

Lederman, Abd-El-Khalick, Bell and Schwartz (2002), in order to establish the construct validity of their Views of Nature of Science Questionnaire, Form B (VNOS-B), have reviewed literature on history, philosophy, epistemology and sociology of science and suggested the importance of the following topics for school science: (1) empirical nature of scientific knowledge; (2) observation, inference and theoretical entities in science; (3) distinctions and relationship between scientific theories and law; (4) creative and imaginative nature of scientific knowledge; (5) theory-laden nature of scientific knowledge; (6) social and cultural embeddedness of scientific knowledge; (7) myth of the scientific method; and (8) tentative nature of scientific knowledge.

In spite of some differences, especially with respect to the terminology used, both groups of researchers (Osborne et al., 2003; Lederman et al., 2002) would recommend about the same set of "ideas-about-science" for inclusion in the science classroom. Nevertheless, there is an important and crucial difference with respect to what Osborne et al. (2003) refer to as "Scientific Method and Critical Testing" and Lederman et al. (2002) as "Myth of the Scientific Method". It is important to note that most science curricula and textbooks in most parts of the world emphasize inductivism, falsificationism and the scientific method. Given this perspective, Osborne et al.'s (2003) combining of the "Scientific Method" and "Critical Testing" obscures the issue and even suggests that the use of the "Scientific Method" is perhaps inevitable. In contrast, Lederman et al. (2002) have clearly traced its origin to Francis Bacon's *Novum Organum*, and its unhealthy influence in school science:

> The myth of the scientific method is regularly manifested in the belief that there is a recipelike stepwise procedure that all scientists follow when they do science. This notion was explicitly debunked: There is no single scientific method that would guarantee the development of infallible knowledge (AAAS, 1993; Bauer, 1994; Feyerabend, 1993; NRC, 1996; Shapin, 1996). It is true that scientists observe, compare, measure, test, speculate, hypothesize, create ideas and conceptual tools, and construct theories and explanations. However, there is no single sequence of activities (prescribed or otherwise) that will unerringly lead them to functional or valid solutions or answers, let alone certain or true knowledge.
>
> (Lederman et al., 2002, pp. 501–502)

At this stage, it is important to consider the advice of philosopher-physicist James Cushing (1989), "One cannot simply amass a 'Baconian' heap of facts (or of theses) and then hope that truth or a theory will thereby emerge" (pp. 19–20).

Research Questions, Objectives and Framework of the Study

This study raises the following research questions:

1. What is the effect of the instructional intervention undertaken in this study on teachers' conceptions of nature of science?
2. What are participant teachers' conceptions of what "ideas-about-science" can be included in the classroom?

Based on these research questions the objectives of this study are to facilitate in-service chemistry teachers' understanding of nature of science and what "ideas-about-science" can be included in the classroom, based on the following framework (here nature of science is presented in the following four points):

Point 1: Construction of scientific theories is a complex and creative process, which is not necessarily facilitated by the accumulation of experimental observations.

Point 2: Inclusion of nature of science in the classroom is necessary, in order to facilitate the formation of citizens with critical abilities for development and progress of society.

Point 3: Critical evaluation of the nine "ideas-about-science" that could be included in school science, as suggested by Osborne et al. (2003).

Point 4: Understanding of the scientific method beyond that of a step-by-step recipe can help the resolution of the contradiction between the recommendations of Osborne et al. (2003) and Lederman et al. (2002).

Rationale and Design of the Study

This study is based on 17 in-service chemistry teachers who had enrolled in the course Epistemology of Science Teaching, as part of a Master's degree program in education at a major university in Latin America. Nine teachers worked in secondary schools and eight at the university level (male=6, female 11; age range: 25–45 years), and their teaching experience varied from about 5 to 20 years. In the previous year all teachers had enrolled in the following courses: (a) Methodology of Investigation, in which basic philosophical ideas of Popper, Kuhn and Lakatos were discussed, in order to provide an overview of the controversial nature of progress in science (growth of knowledge) and its implications for research methodology in education. Teachers were familiar with the notion that basic ideas like the scientific method, objectivity and inductive nature of science were considered to be controversial and questionable by philosophers of science; (b) Investigation in the Teaching of Chemistry, in which students' alternative conceptions and conceptual change strategies were discussed within an HPS perspective, with particular reference to historical controversies.

Course Content (Reading List)

In order to provide an overview of the course content, participants were provided the article by Gallegos (1999). The course itself was based on 17 readings and was subdivided into the following sections:

Unit 1: *Nature of Science*: 1. Campanario (1999); 2. Mellado (2003); 3. Matthews (2004); 4. Smith and Scharmann (1999); 5a. Niaz (2001a); 5b. Scharmann and Smith (2001); 6. Fernández, Carrascosa, Cachapuz and Praia (2002).

Unit 2: *Critical Evaluation of Nature of Science*: 7. Blanco and Niaz (1997); 8. Petrucci and Dibar (2001); 9. Osborne, Collins, Ratcliffe, Millar and Duschl (2003); 10. Lin and Chen (2002); 11. Lederman, Abd-El-Khalick, Bell and Schwartz (2002).

Unit 3: *Critical Evaluation of Constructivism*: 12. Moreno and Waldegg (1998); 13. Marín, Solano and Jiménez (1999); 14. Martínez (1999); 15. Gil-Pérez, et al. (2002); 16. Niaz et al. (2003); 17. Abd-El-Khalick et al. (2004).

Course Organization and Activities

On the first day of class (2 hours) all participants were provided copies of all the readings and salient features of the course were discussed. It was emphasized that the course called for active participation. As all teachers worked in nearby schools and universities, three types of course activities were programmed:

1. Class discussions were planned on Saturdays of the 6th and 8th week of the course (3 hours in the morning and 3 in the afternoon). Readings 1–9 were

discussed in the first meeting and readings 10–17 in the second meeting. Teachers were supposed to have studied each of the readings before the meetings. Each meeting started off with various questions and comments by the participants. The instructor intervened to facilitate understanding of the issues involved. (Total time devoted to class discussions = 12 hours.)

2. Class presentations by the participants were programmed during the 11th and final week of the course (Monday to Saturday, total time = 44 hours). On the first day of class, all participants selected (by a draw) one of the 17 readings for a presentation. Each participant was assigned 90 minutes (30 minutes for the presentation and 60 minutes for interventions and discussions). Each of the presentations was moderated by one of the participants. The instructor intervened when a deadlock was reached on an issue. It was expected that the participants would present the important aspects of the readings, with the objective of generating critical discussions. All participants prepared PowerPoint presentations.
3. During class presentations in the final week participants were encouraged to ask their questions in writing, which were then read out loud by the moderator. The presenters were given the opportunity to respond and then a general discussion followed. At the end of each session all written questions were submitted to the instructor, which provided important feedback with respect to issues, conflicts and interests of the participants. Each participant signed her/his question. The same procedure was followed in all 17 presentations, generating a considerable amount of data (a pool of 345 questions). Besides this data, the instructor also took class notes throughout the course. This data facilitated the corroboration (triangulation of data sources) of participants' responses to the four exam questions in this study.

Course Evaluation

All participants presented an Initial exam in the first session and a Final exam during the last session of the final week (11th). There was no particular reason for labeling these evaluations as "Initial" or "Final", except for the fact that university regulations used these labels. However, the rationale behind these evaluations in this study (could have been named differently) was not merely evaluation in the traditional sense. Both exams were "open book" (about 3 hours each) and participants were allowed to consult any material that they felt could be helpful. Initial exam reflected participants' experience based on 12 hours of class discussions during the first 10 weeks. The Final exam reflected an additional experience of 36 hours of class presentations and discussions during the final week. Furthermore, both exams were based on the premise that oral presentations and discussions can be complemented by written responses and thus facilitate greater understanding. This study is based on participants' written responses to Items 1 and 2 of the Initial exam and Items 1 and 3 of the Final exam. These four items were formulated in order to provide possible answers to the four points in Objectives and the Framework section.

Multiple Data Sources

Based on the different course activities this study generated the following data sources:

a. Question–answer sessions after each of the 17 formal presentations, in which participants wrote their questions/comments (interventions), which were then discussed in class and later submitted to the instructor (a pool of 345 questions and comments was generated). *Note*: Proportion of assessment attributed to these activities = 40%.
b. Initial and Final exams during the final week, separated by 36 hours of class presentations and discussions (total time = 8 hours). *Note*: Proportion of assessment attributed to Initial exam = 20%, and Final exam = 20%.
c. Instructor's class notes, based on the following activities throughout the course: class discussions during the 6th and 8th week of the course (total time = 12 hours), 17 formal presentations during the final week, question–answer sessions after each presentation.
d. Critical essay. As part of their evaluation all participants were required to submit a critical essay based on any one or various readings. The objective of this essay was to present a critique based on epistemological, philosophical and methodological aspects. This essay was submitted 10 days after having finished the Final exam, which provided the participants ample time to reflect and elaborate their ideas. *Note*: Proportion of assessment attributed to this activity = 20%.

At this stage it is important to emphasize the role played by multiple data sources in educational research (*Note*: teachers in this study participated in 56 hours of various class activities). Given the nature of the paradigm wars (Gage, 1989; Howe, 1988), educational literature has suggested the need to move beyond the quantitative versus qualitative research designs and called for mixed methods research, namely, "researchers should collect multiple data using different strategies, approaches and methods in such a way that the resulting mixture or combination is likely to result in complementary strengths" (Johnson & Onwuegbuzie, 2004, p. 18). Recent research considers the mixed methods research as a new and emerging paradigm (Johnson & Christensen, 2004; Johnson & Turner, 2003; Onwuegbuzie & Leech, 2005; Sale & Brazil, 2004). Qualitative researchers have also generally endorsed triangulation of data sources and Guba and Lincoln (1989) consider that: "triangulation should be thought of as referring to *cross-checking specific data items* of a factual nature (number of target persons served, number of children enrolled in a school-lunch program...)" (p. 24, emphasis added). More recently, Guba and Lincoln (2005) have clarified that although qualitative and quantitative paradigms are not commensurable at the philosophical level (i.e., basic belief system or worldview) still, "within each paradigm, mixed methodologies (strategies) may make perfectly good sense" (p. 200). Interestingly, however, Shulman (1986) had advocated the need for hybrid research designs much earlier:

"These hybrid designs, which mix experiment with ethnography, multiple regressions with multiple case studies, process-product designs with analyses of student mediation, surveys with personal diaries, are exciting new developments in the study of teaching" (p. 4). Within this perspective it is plausible to suggest that this study has a hybrid research design. Research reported here is based on actual classroom practice and participants did not feel constrained by the research design. All the data generated (56 hours of class activities) was based on regular classroom activities. An example of a similar study is provided by Niaz (2004a).

Point 1 of Framework: Construction of Scientific Theories is a Complex and Creative Process

This section is based on participants' written responses to the following exam question (Item 1, Initial exam):

> In the context of the scientific contributions of Newton, Darwin, Maxwell and Einstein, Gallegos (1996) has pointed out that, "All the scientists of a given epoch have the same experimental data, and not all of them are able to bring into focus, restructure or reinterpret in the same manner" (p. 324). Based on this statement, which of the following theses do you prefer?
>
> a. There has to be a genius in order to interpret and understand the data.
> b. Science does not advance by just having the data.
> c. Once a scientist has the data it is relatively easy to elaborate a theory.

This question provided the participants an opportunity to reflect, understand and have an opinion with respect to what researchers in science education have referred to as creativity, imagination, diversity of scientific thinking, theory-laden nature of scientific knowledge, analysis and interpretation of data (Lederman et al., 2002; Osborne et al., 2003).

Only one participant selected thesis (a), namely, there has to be a genius in order to interpret and understand the data, and justified it in the following terms:

> scientists tend to interpret data in different ways as they belong to different scientific communities, which grow in different conditions and have different heuristic principles … and it is precisely for this reason that we need a genius in order to reach consensus.

Eleven participants selected thesis (b), namely, science does not advance by just having the data, and following are some of the justifications provided:

> …what is truly important for the advancement of science is the confrontation of ideas, discussions and controversies that can generate interpretations and reinterpretations of data.

> ...according to Lakatos, rival theories can explain the same set of data ... a scientist can interpret data according to her/his personal constructs or heuristics ... nevertheless, it is the scientific community that can facilitate consensus.

> ...in the case of the oil drop experiment, two scientists (Millikan and Ehrenhaft) had the same data and still their interpretations were entirely different. It was the scientific community that decided in favor of Millikan's interpretation. [Millikan–Ehrenhaft controversy had been discussed in one of the courses during the previous year, cf., Holton, 1978a; Niaz, 2000a.]

> ...data tend to impose an "imaginary limit" ... science progresses, precisely because scientists can foresee beyond this "limit" based on the data ... in research on sub-atomic particles the phenomena under study are frequently beyond the limit of detection of the apparatus.

Two of the participants selected thesis (c) and one was partly in agreement with both thesis (b) and (c). Two participants did not agree with any of the three theses, and one of them reasoned, "the three thesis present science as a body of knowledge formed by data and theories, whereas the role of controversies and heuristic principles is ignored".

In order to provide reliability of results (triangulation of data sources), participants' critical essays and interventions during the 17 class presentations were checked (see pool of questions in the section Multiple Data Sources). It was found that 14 participants who selected thesis (a), (b) and (c) expressed similar ideas on at least two different occasions. It is important to note that Guba and Lincoln (1989) consider such "cross-checking specific data items of a factual nature" (p. 241) as an essential part of triangulation of data sources.

Discussion

Participants' responses show that a majority (11 out of 17; 65%) clearly favored thesis (b), namely, science does not advance by just having the data. Participants reasoned quite cogently that progress in science inevitably leads to controversies and alternative interpretations of data. One participant, by invoking the idea of an "imaginary limit", came quite close to what philosophers of science have referred to as under-determination of scientific theories by experimental data, that is the Duhem–Quine thesis (cf. Quine, 1953). Furthermore, according to Lakatos (1999), "Inductivism claims that a proposition is scientific if provable from facts; what we shall now set out to do is to show that no proposition whatsoever can be proven from facts. And certainly not any scientific proposition" (p. 36).

It appears that this course provided participants an opportunity to acquire knowledge with respect to Point 1 of the Framework, which agrees to a fair degree with both Lederman et al. (2002) and Osborne et al. (2003).

Point 2 of Framework: Inclusion of Nature of Science for Development and Progress of Society

This section is based on participants' written responses to the following exam question (Item 2, Initial exam):

> Collins (2000) has presented a trilemma with respect to teaching science due to the following conflicting requirements (reproduced in Reading 9, Osborne et al., 2003, p. 694):
>
> a. The possibility offered by science to discover and create new knowledge.
> b. The dogmatic and authoritarian way of teaching science, based partially on Kuhn's (1962) normal science.
> c. The necessity to teach nature of science in order to appreciate and understand the different aspects of scientific development.
>
> What strategy can you suggest in order to resolve this trilemma?

This exam question provided the opportunity to deal with the horns of a trilemma: On the one hand, school science generally tries to inculcate a dogmatic and authoritarian approach (this is known to be true and so you must learn), and on the other, science also presents a popular culture that promotes emancipation based on scientific discoveries. In this context, it is worthwhile to make teachers and curricula more conscious of how the inclusion of nature of science in the classroom will be resisted and even perhaps found contradictory. Based on participants' responses, the following categories were created:

1. Normal science and the scientific method as an obstacle (n = 7). Seven participants explicitly stated that Kuhn's (1962) normal science manifests itself in the science curriculum and the textbooks through the scientific method. These participants considered this aspect to be the major obstacle in the resolution of the trilemma, and following are examples of some of the responses:

   > ...science in the classroom is presented from the positivist perspective in which the scientific method dominates the scenario—this is what defines science. Similarly, only normal science is taught, as this is what appears in the textbooks and has consensus. This in itself creates a big problem by forcing students to memorize and repeat without understanding what science is all about.

   > ...of the different ideas that can be included in school science, it is the tentative nature of science that can help most in undermining the influence of Kuhn's normal science.

   > ...I suggest eliminating the second horn of the trilemma, that is science cannot be taught as suggested by Kuhn's (1962) normal science, with no

reference to the problems and controversies. It is precisely due to this that school science has so many distortions of what real science is.

2. Trilemma does not necessarily lead to a confrontation (n=6). Six participants argued explicitly that the three horns of the trilemma can be reconciled so that there is no drastic change in classroom activities, and the following are two of the examples:

 > ...in order to resolve the trilemma, I suggest that we invoke Kuhn himself, who suggested that scientists working in different paradigms see things differently. If Kuhn's paradigms could be made flexible and open, then a whole range of activities can be included in the classroom, including nature of science ... this opens the possibility of epistemological pluralism.

 > ...the three aspects of the trilemma can be considered as different paradigms. To resolve the trilemma, we can consider the Lakatosian thesis, namely, the three paradigms could coexist, so that each one of them can complement the other.

3. Introduction of the history and philosophy of science in the classroom can facilitate a new paradigm (n=5). Five participants suggested that resolution of the trilemma is a long-term process, which can be initiated by the use of HPS. This, of course, requires careful reconstruction of the different topics of the science curriculum at different levels of detail and sophistication (secondary/university). The teacher by "unfolding" the different historical episodes can emphasize and illustrate how science actually works (tentative, controversial, rivalries, alternative interpretations), and this will show to the students that the "normal" science presented in their textbooks is in most cases a fiction. The following is an example of the justification presented by a participant: "we need to teach science not just from the empirical perspective but also the heuristics. Students need to have the 'tools' necessary for understanding science ... how scientists elaborate theories—their weak points, and their strongholds."
4. NOS courses for teachers (n=4). Four teachers noted the need for in-service and pre-service courses for teachers based on different aspects of NOS. It was emphasized that at university level neither the science graduates nor those who have degrees in education have been introduced to history, philosophy and epistemology of science. (This, of course, would vary from one country to another.)

The number of responses to this exam question was more than 17 as some participants suggested more than one strategy. In order to provide reliability of results (triangulation of data sources) participants' critical essays and interventions during the 17 class presentations were checked (see pool of questions in the section Multiple Data Sources). It was found that all participants expressed similar ideas on at least two different occasions.

Discussion

Based on the trilemma presented by Collins (2000), it appears that a majority of the participants considered that Kuhn's normal science wields considerable influence in the classroom and hence was the major obstacle for the introduction of new ideas about how science works. Any change would perhaps require the elaboration of a new paradigm based on history, philosophy and epistemology of science, which in the long run could show to the students that the normal science presented in their textbooks is in most cases quite different from what science is all about.

Point 3 of Framework: Critical Evaluation of the Nine "Ideas-about-Science" as Suggested by Osborne et al. (2003)

This section is based on participants' written responses to the following exam question (Item 1, Final exam):

> Osborne et al. (2003, pp. 706–709) have suggested the inclusion in the science curriculum of nine aspects based on nature of science:
>
> a. Select three aspects that in your opinion are the most important and order them from less to more important, based on a scale of 1–5 points;
> b. Do you consider that it is necessary to include (not suggested) or exclude some of those aspects suggested by these authors?

At this stage (Final exam) participants had read varied points of view in the 17 readings, discussed, made formal presentations, criticized their classmates and the original authors and presented the Initial exam over a period of 11 weeks that required 54 hours of direct contact in the classroom. It was expected that this experience could enable this group of 17 in-service teachers to evaluate critically the nine "ideas-about-science" suggested by 23 experts in the study by Osborne et al. (2003). In comparison to the experts this group of teachers formed a fairly homogeneous group. Similar to the experts, these teachers were also asked to rate the importance of the nine aspects on a scale of 1–5 points.

Discussion

a. Selection of three aspects and the ratings of the teachers. Table 10.1 shows that the four most important aspects for this group of chemistry teachers were (based on mean score): creativity (4.50), historical development of scientific knowledge (4.29), diversity of scientific thinking (4.17) and scientific method and critical testing (3.83). Nevertheless, it is important to note that historical development of scientific thinking was selected by 14 teachers. Interestingly, none of the teachers selected one of the aspects (hypothesis and prediction). Table 10.1 also shows that two participants gave the maximum importance (5 points) to aspect 1 (scientific method and critical

Table 10.1 Critical Evaluation of the Nine Aspects (Ideas-About-Science) by Chemistry Teachers Based on Point 3 of Framework (Exam Question a)

Ideas-About-Science Suggested by experts in Osborne et al. (2003)	*Chemistry Teachers' Ratings*	*n*	*Mean Score*
1. Scientific method and critical testing	5, 4, 5, 4, 2, 3	6	3.83
2. Creativity	5, 5, 4, 4	4	4.50
3. Historical development of scientific knowledge	3, 2, 5, 5, 4, 5, 3, 5, 5, 5, 5, 5, 4, 4	14	4.29
4. Science and questioning	1, 3, 4, 3, 2	5	2.60
5. Diversity of scientific thinking	4, 4, 4, 4, 5, 4	6	4.17
6. Analysis and interpretation of data	4, 3, 2, 3, 3, 1, 2	7	2.57
7. Science and certainty	3, 3, 4, 5, 3, 4, 3	7	3.57
8. Hypothesis and prediction	–	–	–
9. Cooperation and collaboration in the development of scientific knowledge	4, 1	2	2.50

Notes
1. Chemistry teachers' ratings in this study are based on a scale of 1–5 points, 5 representing most important. The same scale was used by the experts in Osborne et al. (2003).
2. Each teacher was asked to select three aspects. Ratings represent the evaluation (1–5 points) of each teacher on a particular aspect. There are a total of 51 (17 × 3) ratings.
3. n represents the number times an aspect was selected.
4. Mean score is based on ratings and the number of times it was selected.

testing). Both these participants argued that they did not conceive the scientific method in the traditional sense as a rigid sequence of steps, but rather selected this aspect as it was accompanied by "critical testing", which meant "hypotheses, predictions, data, analyses and interpretations, all this accompanied by creativity of the researchers". It is important to note that one of the experts in Osborne et al. (2003) also came quite close to this understanding, "in the research world of science, careful experimentation is used to test hypotheses" (p. 706). It seems that "scientific method" and "critical testing" should constitute two different aspects of NOS. This also explains why none of the participants selected hypothesis and prediction, as they perhaps considered it to be an integral part of critical testing (already included in aspect 1). *Note*: In order to provide reliability of results (triangulation of data sources) for this section, participants' critical essays and interventions during the 17 class presentations were checked (see pool of questions in the section Multiple Data Sources). It was found that all participants expressed similar ideas on at least two different occasions.

b. Inclusion or exclusion of some aspects. The second part of this research question asked participants to suggest some aspects that Osborne et al.

(2003) may not have suggested that they would like to include. Participants suggested the following aspects: (i) science and technology (n=2); (ii) role of communication and "invisible colleges" (n=2); (iii) presuppositions of the scientists (n=2); (iv) ethical and moral issues and fraud (n=3); (v) none (n=9). Some of these aspects came up for discussion in the previous courses and Readings 1, 5a and 6. Participants were also asked if they would like to exclude some of the aspects suggested by Osborne et al. (2003) and the following are the results: aspect 1 (scientific method and critical testing), n=5; aspect 4 (science and questioning), n=1; aspect 5 (diversity of scientific thinking), n=1; aspect 7 (science and certainty), n=3; aspect 9 (cooperation and collaboration…), n=2; none, n=8. Once again, it seems that scientific method and critical testing (aspect 1) was controversial, as five participants suggested that it be excluded.

Point 4 of Framework: Understanding Scientific Method and Resolution of the Contradiction between Lederman et al. (2002) and Osborne et al. (2003)

This section is based on participants' written responses to the following exam question (Item 3, Final exam):

> What do you understand by the scientific method? According to Reading 9 (Osborne et al., 2003, p. 706) it is the "central thrust of scientific research", whereas Reading 11 (Lederman et al., 2002, p. 501) considers that a "recipe-like stepwise procedure" is a myth. How do you resolve this contradiction between these two groups of researchers?

This exam question provided the opportunity for not only resolving the contradiction between two groups of researchers but also facilitating greater understanding with respect to an important issue. Most chemistry and methodology textbooks still present the scientific method as a recipe-like stepwise procedure. Based on participants' responses, the following categories were created:

1. A transaction between the positions of Lederman et al. (2002) and Osborne et al. (2003), n=7. These participants argued that the two positions can be reconciled and the following is an example:

> Discussion of historical episodes in the classroom can illustrate how scientists do follow certain procedures, for example construction of atomic models by Thomson and Rutherford. However, a stage comes when the procedures break down and a controversy ensues—this is what happened in the case of atomic models. If there had been an infallible scientific method, both Thomson and Rutherford could have met over dinner and resolved the controversy … In conclusion, science has its methods of investigation (observe, compare, measure, experiment, hypothesize, interpret)…, however, these

procedures cannot be followed sequentially ... [This response refers to the Thomson–Rutherford controversy that had been discussed in a course in the previous year, cf. Niaz, 1998.]

2. In favor of Osborne et al.'s (2003) position, n = 1. One participant supported this position and reasoned in the following terms:

> all scientific activity is sustained by different processes that orient in a coherent and organized way ... why not accept the scientific method if its presence has been innate in all research over time and will continue to be so.

It is interesting to follow the line of reasoning of this participant through this course: Exam question with respect to Point 1 of the Framework, response (c), namely, once a scientist has the data it is relatively easy to elaborate a theory (comes quite close to an empiricist perspective); Exam question with respect to Point 2 of the Framework, suggested that the scientific method needs to be reconceptualized so that it does not lead to dogmatic teaching of science; Exam question (a) with respect to Point 3 of the Framework, selected aspect 1 (scientific method and critical testing) with a rating of 4 points; Exam question (b) with respect to Point 3 of the Framework, did not suggest the inclusion of any new aspect but excluded aspect 7 (science and certainty). This constitutes a fairly coherent position, in which the participant recognized the importance of the scientific method, which might need some reconceptualization for educational purposes.

3. In favor of Lederman et al.'s (2002) position, n = 9. This group of participants was quite concerned with respect to this contradiction, especially as both studies were conducted by science educators and published in the same journal. Two of the participants specifically suggested that the two groups of science educators, namely, Osborne et al. (2003) and Lederman et al. (2002) owe an explanation to the community with respect to their different ways of emphasizing the scientific method in the classroom. The following are two examples that represent the views of this group of participants:

> ...history of science contains many episodes in which the scientific method did not work ... e.g., Einstein's theory of relativity, Millikan's determination of the elementary electrical charge, formulation of atomic models by Thomson, Rutherford and Bohr. It is surprising that Osborne et al. (2003) seem to agree with their panel of experts, namely, that the scientific method constitutes the "central thrust of scientific research." At best one could accept that the scientific method is one among other strategies used by scientists. History of science provides a good antidote for demolishing the myth of a recipelike procedure. [These episodes from history of science had been discussed in courses during the previous year, cf. Niaz, 1998, 2000a.]

> ...it is true that in every scientific investigation there are observations, inferences, interpretations, hypotheses etc ... these processes cannot be systematized, and this is precisely why the scientific method is being criticized. It is

important that research in nature of science must be conducted by the community of science educators ... the panel of experts in Osborne et al. (2003) perhaps was not entirely aware of our needs.

Discussion

This exam question highlights some of the difficulties faced by participants with respect to how the scientific method could be included in the classroom. One of the participants considered this to be fairly simple and thus was in agreement with the panel of experts in Osborne et al. (2003). Seven of the participants pointed out difficulties but still considered that the two positions could be reconciled. Nine of the participants were openly critical of the present stage of research with respect to the scientific method and suggested that the two groups of researchers (Lederman et al., 2002 and Osborne et al., 2003) need to provide further guidance on this crucial issue. Interestingly, five participants had suggested the exclusion of aspect 1 (scientific method and critical testing) on Exam question (b) with respect to Point 3 of the Framework, and these participants also favored the position of Lederman et al. (2002) on this exam question (Point 4 of the Framework), which shows the consistency of these participants' thinking.

Note: In order to provide reliability of results (triangulation of data sources) for this exam question, participants' critical essays and interventions during the 17 class presentations were checked (see pool of questions in the section Multiple Data Sources). It was found that all participants expressed similar ideas on at least two different occasions.

Conclusions and Educational Implications

The group of in-service chemistry teachers who participated in this study had a fair amount of experience in their courses during the previous year with respect to the scientific method, positivism, objectivity, inductive nature of science and historical controversies within an HPS perspective. For this reason this group was particularly interested in and apprehensive of what "ideas-about-science" could be included in the classroom. Experts in the Osborne et al. (2003), Delphi-style study assigned the highest priority to aspect 1 (scientific method and critical testing) and considered it to be the "central thrust of scientific research" (p. 706). Furthermore, Osborne et al. (2003, table 4, p. 713) also consider aspect 1 to be comparable to "scientists require replicability and truthful reporting" in McComas and Olson (1998). This comparison at best is an overstatement, as critical testing is quite different from replicability. Combination of "scientific method" and "critical testing" in the same category is also quite problematic, as some teachers agreed with critical testing and still did not feel the necessity of invoking the scientific method. It can be argued that Osborne et al. (2003) and the experts also may not endorse that there is a recipe-like stepwise procedure in science. If this were so then it is all the more difficult to understand why the scientific method was included in aspect 1, as critical testing by itself could constitute an aspect. There are other aspects in the study that can easily represent the processes of science, namely, analysis and interpretation of data (aspect 6) and hypothesis and prediction (aspect 8). Apparently, the experts wanted to not only emphasize the processes of science but rather go beyond.

Based on these considerations and participants' responses to the four exam questions, this study has the following educational implications:

1. It is important for teachers to understand that science does not advance by just doing the experiments and having the data. Progress in science inevitably leads to controversies and alternative interpretations of data. This task is difficult to accomplish as most science curricula, textbooks and teachers present science as "normal science" (Kuhn, 1962), which is different from what science is all about. This makes the question, "What ideas about science should be taught in school science?" all the more pertinent. It is rather paradoxical that progress in science is generally considered to promote emancipation and still school science inculcates a dogmatic and authoritarian approach.
2. Experimental data tend to impose an "imaginary limit" on our abilities and science progresses precisely because scientists can foresee this "limit" based on the data (based on exam question with respect to Point 1 of the Framework). This constitutes what philosophers of science have referred to as under-determination of scientific theories by experimental data (Quine, 1953) and also as idealization (McMullin, 1985). This is a fairly novel way of understanding this aspect of the dynamics of scientific progress (cf. Niaz, 2009a).
3. Kuhn's "normal science" manifests itself in the science curriculum and textbooks through the scientific method and wields considerable influence (based on exam question with respect to Point 2 of the Framework). Given teachers' criticism of dogmatic and authoritarian ways of teaching science, the concern with respect to the scientific method is quite understandable.
4. The trilemma posed by Collins (2000) can be resolved by introducing history, philosophy and epistemology of science, which in the long run could convince students (and society) that normal science presented in the curriculum and textbooks is quite different from what science is all about (based on exam question with respect to Point 2 of the Framework).
5. Of the nine aspects endorsed by experts in Osborne et al. (2003), chemistry teachers in this study considered the following to be the most important: creativity, historical development of scientific knowledge, diversity of scientific thinking, and scientific method and critical testing (based on exam question with respect to Point 3 of the Framework). In spite of serious reservations with respect to the scientific method, it seems that some teachers were swayed by the opinion of the experts and its publication in a leading journal of science education.
6. Scientific method and critical testing (aspect 1 in Osborne et al., 2003) was considered to be not only controversial but also somewhat ambiguous

(based on exam questions with respect to Points 3 and 4 of the Framework). It was suggested that the two could be considered as separate aspects.

7. With respect to the contradiction between the positions of Lederman et al. (2002) and Osborne et al. (2003), only one participant supported the latter, namely, inclusion of scientific method, aspect 1 (based on exam question with respect to Point 4 of the Framework). Nine participants supported the position of Lederman et al. (2002), that is, recognized the scientific method as a myth and suggested the inclusion of history of science in order to show its pitfalls. Interestingly, seven participants considered that the two positions could be reconciled.
8. Participants were critical of the present stage of research with respect to the scientific method and suggested further research on this crucial issue (based on exam question with respect to Point 4 of the Framework).

Finally, it is important to note that Cushing's (1989) advice on this subject can be helpful to understand the issues involved: "science is an historical entity whose practice, methods and goals are *contingent*. There may not be *a* rationality which is the hallmark or the essence of science" (p. 2, original italics. In a footnote the author explains what he means by *contingent*, "I simply mean not fixed by logic or necessity", p. 20). According to Windschitl, Thompson and Braaten (2008), 100 years after its conception, the scientific method continues to reinforce a kind of cultural lore about what it means to participate in inquiry. As commonly implemented in places ranging from middle school classrooms to undergraduate laboratories, it emphasizes the testing of predictions rather than ideas, focuses learners on material activity at the expense of deep subject-matter understanding and lacks epistemic framing relevant to the discipline.

Next Chapter

This chapter discussed the difficulties involved in introducing NOS in the classroom, primarily due to popularity of the scientific method and its relationship with Kuhn's "normal science". The objective of Chapter 11 was to familiarize teachers with one important aspect of NOS, namely the tentative nature of scientific knowledge. The rivalry between different forms of constructivism in science education was considered to be a manifestation of the tentative nature of scientific knowledge. Furthermore, an analogy was drawn with various atomic models in the history of science that competed with each other. Consequently, most teachers understood that it is precisely the contradictions faced by existing forms of constructivism that would facilitate alternative and rival theories, leading to new advances. This may require more eclectic and perhaps pragmatic approaches. Finally, based on an exploration of the tentative nature of scientific knowledge and its critical appraisal, this study facilitated a better understanding of the current status of constructivism in science education.

Chapter 11

Whither Constructivism?

Understanding the Tentative Nature of Scientific Knowledge*

Introduction

The decline of positivism during the latter half of the 20th century facilitated the development of constructivism in various forms as an alternative philosophical and educational theory (Louden & Wallace, 1994). Most science educators would agree that during the 1970s and the 1980s, among other forms of constructivism Piagetian and Ausubelian constructivism played a dominant role. Piagetian constructivism emphasized the need for going beyond expository teaching practice in order to facilitate development of reasoning based on the learning cycle. In contrast, Ausubelian constructivism promoted meaningful receptive learning based on prior knowledge of the students and concept maps. Since then, constructivism in science education has developed in many forms by drawing inspiration from various philosophical and epistemological sources (Geelan, 1997; Good, 1993; Phillips, 1995). Of the different forms, radical (Glasersfeld, 1989) and social constructivism (Glasson & Lalik, 1993; Tobin & LaMaster, 1995) have enjoyed more popularity with science educators. For radical and some social constructivists, experience is the ultimate arbiter for deciding between scientific theories and how students acquire knowledge. Despite the popularity, almost all forms of constructivism have also been the subject of scrutiny and critical appraisal (De Berg, 2006; Geelan, 2006; Kelly, 1997; Matthews, 1993; Niaz, 2001d; Osborne, 1996; Phillips, 1995; Solomon, 1994; Suchting, 1992; Taber, 2006).

An important aspect of the development of different forms of constructivism in science education is the need for a continual critical appraisal. Early debates (Novak, 1977) provided the stimulus for this continued progressive development. More recently, Nola (1997) has emphasized that popular forms of constructivism (radical and social) will have to compete and often unfavorably with rival views. Competition between rival theories, tentative nature of science and theory-ladenness of observations are important contributions of the new philosophy of science, which has permeated science education research (Lederman, Abd-El-Khalick, Bell & Schwartz, 2002). Tsai (2006) has emphasized the importance of these aspects of nature of science for constructivism and teacher

* Reproduced with permission from: Niaz, M. (2008e). Whither constructivism? A chemistry teachers' perspective. *Teaching and Teacher Education, 24*, 400–416.

training programs. Table 11.1 provides an outline of the tentative nature of science in the two domains, namely, atomic structure and constructivism in science education.

At this stage it is important to note that the different forms of constructivism in science education have as much to do with the different psychological models of teaching and learning (developmental stage theory, socio-cultural, motivational perspectives, etc.) as with different views on scientific epistemology. Kitchener (1986) an important scholar on genetic epistemology has expressed this in cogent terms, "Piaget attempts to explain the growth of knowledge as Popper and Lakatos do, by providing a rational reconstruction of the course of epistemic change in which transitions occur by virtue of certain normative principles" (p. 210). Similarly, Pascual-Leone (1987), a leading neo-Piagetian psychologist, has emphasized the constructive perspective which:

> presupposes that subjects construct their own world of experience (objects, events, transformations) by means of cognitive structures and organismic regulations/factors. This constructed world, however, is valid only if it epistemologically *reflects* distal objects, distal events and transformations actually occurring in the environment.
>
> (p. 534, original italics)

Table 11.1 Tentative Nature of Scientific Theories

Atomic Structure	*Constructivism in Science Education*
1897 Thomson	1960 Trivial Constructivism (Piaget)
1911 Rutherford	1970 Human Constructivism (Ausubel, Novak)
1913 Millikan	
1913 Bohr	1980 Radical Constructivism (Glasersfeld)
1916 Sommerfeld	
1924 De Broglie	1990 Social Constructivism (Vygotsky)
1925 Pauli	
1925 Heisenberg	1999 Pragmatic Constructivism (Perkins)
1926 Schrödinger	
1932 Chadwick	
1963 Gell-Mann (Postulating Quarks)	
1997 Perl (Isolating Quarks)	

Notes

1. Under *Atomic Structure* appear the names of prominent scientists who made a significant contribution toward a greater understanding of atomic structure. Inclusion of these names follows a historical sequence (markers), indicating the tentative nature of atomic theories.
2. Under *Constructivism in Science Education* appear the names of prominent psychologists who facilitated a greater understanding of constructivism. Again, although the sequence is historical, various theorists (besides those mentioned) influenced the different forms of constructivism at the same time.
3. Comparison of progress in atomic structure and constructivism in science education draws a parallel between two domains of knowledge.
4. Comparison between the two domains of knowledge was discussed throughout the course and this table is a reconstruction of those discussions.

Niaz (1992, 2005a) has postulated a "progressive problemshift" (Lakatos, 1970) between Piaget's epistemic subject (a general model that neglects individual differences) and Pascual-Leone's metasubject which incorporates a framework for individual difference variables.

More recently, a group of science educators has interpreted the present state of constructivism in science education as, "an impressive development throughout the last two decades" (Gil-Pérez et al., 2002, p. 557) which has facilitated an "emergent consensus" leading to a "paradigm change". This perspective has Kuhnian overtones (periods of "normal science" separated by paradigms) and lacks the understanding that a "consensus" is at most a transitory feature of scientific progress in both science and education. In contrast, Niaz et al. (2003) have argued that constructivism in science education (like any scientific theory) will continue to progress and evolve through continued critical appraisals. Given the popularity in education of the Kuhnian thesis of paradigms being replaced (Lincoln, 1989) and not the coexistence and rivalry among paradigms, it is not surprising that many science educators also follow the same philosophical thesis (Loving & Cobern, 2000). The relationship between Kuhnian philosophy and constructivism in science education has been recognized explicitly by Hodson (1988a): "Kuhnian models of science and scientific practice have a direct equivalent in psychology in the constructivist theories of learning. There is, therefore, a strong case for constructing curriculum along Kuhnian lines" (p. 32). Similarly, Matthews (2004) has traced the historical origins of Kuhn's influence and the acceptance of relativist and anti-realist views by constructivists in science education, namely, as different scientific theories were incommensurable, no rational decision could be made between competing theories. Interestingly, the Kuhnian thesis has also been questioned by a leading educational theorist:

> Where Kuhn erred, I believe, is in diagnosing this characteristic [controversies/conflicts] of the social sciences as a developmental disability ... it is far more likely that for the social sciences and education, the coexistence of competing schools of thought is a natural and quite mature state.
>
> (Shulman, 1986, p. 5)

This shows that Kuhnian theory is not very helpful in understanding constructivism and science education in general. Various critiques have shown that science educators have been overly influenced by Kuhn (cf. Loving & Cobern, 2000; Matthews, 2004). This course took a special interest in making teachers aware of the pitfalls involved in applying Kuhnian theory in science education. It is precisely for this reason that Matthews (2004) was included as a reading material in this course (see Course Content). Furthermore, two of the evaluation items (3 and 6) in this study explicitly dealt with Kuhnian paradigms (see Course Evaluation). To recap, research on constructivism in science education over the last three decades has been a process of competing research programs (Piagetian, Ausubelian, Radical, Social and other forms of constructivism), that has facilitated a constructive dialogue leading to "progressive" transitions (Lakatos, 1970). Among other forms of constructivism, Dialectic constructivism was mentioned which refers to the neo-Piagetian (Pascual-Leone, 1978) attempts to understand construction of knowledge by the students (cf. Niaz, 2001d). It is important to note that due to limitations of time and the course format, this study did not attempt a comprehensive review of research related to constructivism in science education. For a more nuanced understanding of constructivism in science education, see Cobern and Loving (2008).

A review of the literature shows that two previous studies (Glasson & Lalik, 1993; Tobin & LaMaster, 1995) explored the experiences of science teachers as they moved from traditional positivist teaching practices to implementing social constructivism. These two studies formed part of a course in the previous year (cf. Methodology of Investigation, in the section Rationale and Design of the Study), and thus provided the background knowledge for the present study. Consequently, this study goes beyond by presenting teachers a perspective based on the tentative nature of scientific theories and hence the need to understand the different forms of constructivism in science education.

Based on these considerations, the objectives of this study are to:

1. Familiarize chemistry teachers' understanding that different forms of constructivism are a consequence of the proliferation of theories that help us to understand scientific change. (For example, Kuhn's and Lakatos' theories of scientific change provide distinct ways of understanding constructivism.)
2. Facilitate chemistry teachers' understanding that the tentative nature of scientific knowledge leads to the coexistence and rivalries among different forms of constructivism in science education. (In other words, critical appraisals of the different forms of constructivism facilitates the resolution of contradictions and thus leads to further development.)

Rationale and Design of the Study

This study is based on 17 in-service chemistry teachers who had enrolled in the course Epistemology of Science Teaching, as part of a Master's degree program in education at a major university in Latin America. Nine teachers worked in secondary schools and eight at university level (male = 6, female 11; age range: 25–45 years), and their teaching experience varied from about 5 to 20 years. In the previous year all teachers had enrolled in the following courses:

a. *Methodology of Investigation*, in which basic philosophical ideas of Popper, Kuhn and Lakatos were discussed, in order to provide an overview of the controversial nature of progress in science (growth of knowledge) and its implications for research methodology in education. Teachers were familiar with the notion that basic ideas like the scientific method, objectivity and inductive nature of science were considered to be controversial and questionable by philosophers of science. Constructivism was another topic discussed in this course, based on the following readings: Brown (1994), Tobin and LaMaster (1995) and Glasson and Lalik (1993). Brown (1994) critiques Glasersfeld's radical constructivism as it down-plays the role of the teacher.

Based on post-structuralism and hermeneutic phenomenology, an alternative is provided that reasserts the teacher's role. Tobin and LaMaster (1995) and Glasson and Lalik (1993) recount the experiences of two teachers who start out with traditional positivist strategies and finally grapple with the difficulties involved in implementing social constructivism. Comparison of the teaching strategies of Sarah (a teacher in Tobin and LaMaster, 1995) and Martha (a teacher in Glasson and Lalik, 1993) based on social constructivism facilitated participants' understanding of classroom practice.

b. *Investigation in the Teaching of Chemistry*, in which students' alternative conceptions and conceptual change strategies were discussed within a history and philosophy of science perspective, with particular reference to historical controversies. For example, J.J. Thomson did not find E. Rutherford's interpretation of alpha particle experiments as logical, despite chemistry textbooks' claim to the contrary. Similarly, Bohr's postulation of the "quantum of action" was not accepted as a logical alternative by many renowned physicists (cf. Niaz, 1998; this was one of the readings in this course). The Millikan–Ehrenhaft controversy with respect to the oil drop experiment was another controversy discussed in this course. This provided evidence for alternative interpretations of similar experimental data (Duhem–Quine thesis) and was discussed in one of the readings (Niaz, 2000a). This course facilitated participants' understanding with respect to the tentative nature of science and that progress in science inevitably involves rivalries among different interpretations/research programs.

Course Content (Reading List)

In order to provide an overview of the course content, participants were provided the article by Gallegos (1999). The course itself was based on 17 readings and was subdivided in the following sections:

Unit 1: *Nature of Science*: 1. Campanario (1999); 2. Mellado (2003); 3. Matthews (2004); 4. Smith and Scharmann (1999); 5a. Niaz (2001a); 5b. Scharmann and Smith (2001); 6. Fernández, Carrascosa, Cachapuz and Praia (2002).

Unit 2: *Critical Evaluation of Nature of Science*: 7. Blanco and Niaz (1997); 8. Petrucci and Dibar (2001); 9. Osborne, Collins, Ratcliffe, Millar and Duschl (2003); 10. Lin and Chen (2002); 11. Lederman, Abd-El-Khalick, Bell and Schwartz (2002).

Unit 3: *Critical Evaluation of Constructivism*: 12. Moreno and Waldegg (1998); 13. Marín, Solano and Jiménez (1999); 14. Martínez (1999); 15. Gil-Pérez, et al. (2002); 16. Niaz et al. (2003); 17. Abd-El-Khalick et al. (2004).

Course Organization and Activities

On the first day of class (2 hours) all participants were provided copies of all the readings and salient features of the course were discussed. It was emphasized that the course called for active participation. As all teachers worked in nearby schools and universities, three types of course activities were programmed:

1. Class discussions were planned on Saturdays of the 6th and 8th week of the course (3 hours in the morning and 3 in the afternoon). Readings 1–9 were discussed in the first meeting and readings 10–17 in the second meeting. Teachers were supposed to have studied each of the readings before the meetings. Each meeting started with various questions and comments by the participants. The instructor intervened to facilitate understanding of the issues involved. (Total time devoted to class discussions = 12 hours.)
2. Class presentations by the participants were programmed during the 11th and final week of the course (Monday to Saturday, total time = 44 hours). On the first day of class all participants selected (by a draw) one of the 17 readings for a presentation. Each participant was assigned 90 minutes (30 minutes for the presentation and 60 minutes for interventions and discussions). Each of the presentations was moderated by one of the participants. The instructor intervened when a deadlock was reached on an issue. It was expected that the participants would present the important aspects of the readings, with the objective of generating critical discussions. All participants prepared PowerPoint presentations.
3. During class presentations in the final week participants were encouraged to ask their questions in writing, which were then read out loud by the moderator. The presenters were given the opportunity to respond and then a general discussion followed. At the end of each session all written questions were submitted to the instructor, which provided important feedback with respect to issues, conflicts and interests of the participants. Each participant signed her/his question. The same procedure was followed in all 17 presentations, generating a considerable amount of data (a pool of 345 questions and comments). Besides this data, the instructor also took class notes throughout the course. This data facilitated the corroboration (triangulation of data sources) of participants' responses to the seven exam questions in this study.

Course Evaluation

All participants presented an Initial exam in the first session and a Final exam during the last session of the final week. There was no particular reason for labeling these evaluations as "Initial" or "Final", except for the fact that university regulations used these labels. However, the rationale behind these evaluations in this study (which could have been named differently) was not merely evaluation in the traditional sense. Both exams were "open book" (about 3 hours each) and participants were allowed to consult any material that they felt could be helpful. Initial exam reflected participants' experience during the first 10 weeks, including 12 hours of class discussions. The Final exam reflected an additional experience of 36 hours of class presentations and discussions during the final week. Furthermore, both exams were based on the premise that oral presentations and discussions can be complemented by written responses and thus facilitate greater understanding. This study is based on participants' written responses to four items of the Initial exam and three items of the Final exam. These seven items were formulated in order to provide possible answers to the two objectives of this study. The seven items included in the Initial and Final exams are presented below.

Initial Exam

Item 1: What do you understand by constructivism?
Item 2: Why are there so many forms of constructivism?
Item 3: In your opinion, can the present state of constructivism in science education be considered as a paradigm, as conceived by Kuhn (1962)?
Item 4: Which form of constructivism do you prefer?

Final Exam

Item 5: Indicate one aspect of social constructivism that you do not share and suggest an alternative.
Item 6: Given the popularity of social constructivism, do you think it constitutes a paradigm (Kuhn, 1962)?
Item 7: If scientific knowledge is tentative, do you think that the present state of constructivism must also evolve toward other forms?

Items 1, 2, 3 and 4 of the Initial exam refer explicitly to the first objective of the study, namely, various forms of constructivism are a consequence of the proliferation of theories of scientific change. The four items form almost a logical sequence: participants' personal understanding of what is constructivism (Item 1), cognizant of the fact that there are many forms of constructivism (Item 2), constructivism as a Kuhnian paradigm (Item 3) and finally participants' preference for a particular form of constructivism (Item 4). Items 5, 6 and 7 (Final exam) refer explicitly to the second objective of the study, namely, scientific knowledge is tentative, which leads to rivalries and controversies in both science and constructivism. Item 5 explicitly asked the participants to be critical and suggest an alternative. Item 6 explicitly referred to social constructivism as a Kuhnian paradigm and Item 7 suggests the possibility of new forms of constructivism.

Multiple Data Sources

Based on the different course activities this study generated the following data sources:

a. Question–answer sessions after each of the 17 formal presentations, in which participants wrote their questions/comments (interventions), which were then discussed in class and later submitted to the instructor (a pool of 345 questions and comments were generated).

b. Initial and Final exams during the final week, separated by 36 hours of class presentations and discussions (total time = 8 hours).
c. Instructor's class notes, based on the following activities throughout the course: class discussions during the 6th and 8th week of the course (total time = 12 hours), 17 formal presentations during the final week, question-answer sessions after each presentation. These notes consisted of quotes from student (including their names) interventions during presentations and discussions, accompanied by general comments.
d. Critical essay. As part of their evaluation all participants were required to submit a critical essay based on any one or various readings. The objective of the essay was to present a critique based on epistemological, philosophical and methodological aspects. The essay was submitted 10 days after having finished the Final exam, which provided the participants ample time to reflect on and elaborate their ideas.

At this stage it is important to emphasize the role played by multiple data sources in educational research (*Note*: teachers in this study participated in 56 hours of various class activities). Given the nature of the paradigm wars (Gage, 1989; Howe, 1988), educational literature has suggested the need to move beyond the quantitative versus qualitative research designs and called for mixed methods research, namely, "researchers should collect multiple data using different strategies, approaches and methods in such a way that the resulting mixture or combination is likely to result in complementary strengths" (Johnson & Onwuegbuzie, 2004, p. 18). Recent research considers the mixed methods research as a new and emerging paradigm (Johnson & Christensen, 2004; Johnson & Turner, 2003; Onwuegbuzie & Leech, 2005; Sale & Brazil, 2004). Qualitative researchers have also generally endorsed triangulation of data sources and Guba and Lincoln (1989) consider that: "triangulation should be thought of as referring to *cross-checking specific data items* of a factual nature (number of target persons served, number of children enrolled in a school-lunch program…)" (p. 24, emphasis added). More recently, Guba and Lincoln (2005) have clarified that although qualitative and quantitative paradigms are not commensurable at the philosophical level (i.e., basic belief system or worldview) still, "within each paradigm, mixed methodologies (strategies) may make perfectly good sense" (p. 200). Interestingly, however, Shulman (1986) had advocated the need for hybrid research designs much earlier: "These hybrid designs, which mix experiment with ethnography, multiple regressions with multiple case studies, process-product designs with analyses of student mediation, surveys with personal diaries, are exciting new developments in the study of teaching" (p. 4). Within this perspective it is plausible to suggest that this study has a hybrid research design. Research reported here is based on actual classroom practice and participants did not feel constrained by the research design. All the data generated (56 hours of class activities) was based on regular classroom activities. Example of a similar study is provided by Niaz (2004a).

Issues Related to Constructivism Discussed in Class

In order to facilitate participant teachers' understanding of constructivism, various aspects of nature of science, history and philosophy of science and Kuhn's philosophy were discussed in the following readings:

a. Gallegos (1999), critiques inductivism within an epistemological perspective (Popper, Kuhn, Lakatos, Toulmin, Polanyi, Chalmers) and provides a rationale for the different forms of constructivism.
b. Mellado (2003), Reading 2, establishes an analogy between science teachers' conceptual change and various models of progress in science, such as positivism, Popper, Kuhn, Lakatos, Toulmin and Laudan.
c. Matthews (2004), Reading 3, presents a critical appraisal of Kuhn's philosophy of science and draws attention to the uncritical acceptance of his ideas by the science education community, and how this has influenced constructivism.
d. Smith and Scharmann (1999), Reading 4, draw attention to the importance of nature of science for science teachers and suggest the following as essential for understanding progress in science: science is empirical, scientific claims are testable/falsifiable, science is tentative and self-correcting.
e. Niaz (2001a), Reading 5a, critiques Smith and Scharmann (1999) and suggests that following aspects of nature of science also be included: competition among rival theories, same experimental data can be interpreted in more than one way (theory-ladenness of observations) and inconsistent foundation of some scientific theories.
f. Scharmann and Smith (2001), Reading 5b, is a response to Niaz (2001a) and facilitates further understanding of nature of science.
g. Blanco and Niaz (1997), Reading 7, draw attention to the epistemological beliefs of chemistry teachers and students in Venezuela and how that can affect their understanding of progress in science.
h. Petrucci and Dibar (2001), Reading 8, studied understanding of nature of science by students in Argentina.
i. Osborne et al. (2003), Reading 9, refer to the contested nature of science and constructivism within the science education community.
j. Lin and Chen (2002), Reading 10, provide a teaching strategy for promoting preservice chemistry teachers' understanding of nature of science through history.
k. Lederman et al. (2002), Reading 11, provide details of an instrument for assessment of students' conceptions of nature of science.
l. Moreno and Waldegg (1998), Reading 12, present Piagetian constructivism within a perspective of psychogenesis and history of science.
m. Marín, Solano and Jiménez (1999), Reading 13, present a critical appraisal of the various forms of constructivism (social, radical, human, Piagetian).
n. Martínez (1999), Reading 14, presents a critical appraisal of radical constructivism within a philosophy of science perspective.

o. Gil-Pérez et al. (2002), Reading 15, consider the present state of constructivism as that of an emergent consensus that approximates to that of Kuhn's "normal science".
p. Niaz et al. (2003), Reading 16, have critiqued Gil-Pérez et al. (2002) and drawn attention to not a consensus but a continual critical appraisal of constructivism within the science education community.
q. Abd-El-Khalick et al. (2004), Reading 17, discuss issues such as constructivism, nature of science and inquiry learning within a history and philosophy of science perspective in various countries.

This synopsis of the course content shows quite clearly how participant teachers were constantly provided feedback with respect to constructivism and related topics such as nature of science, history and philosophy of science. Furthermore, teachers participated and became aware of the continuing debate on these topics.

Results and Discussion

In this section participants' responses to the seven items are presented and discussed. Participants in this study had discussed various issues related to constructivism in previous courses (cf. section entitled Rationale and Design of the Study), and were quite familiar with the controversial aspects of radical and social constructivism. Furthermore, the previous courses also provided a framework to understand the tentative nature of science based on historical controversies.

Participants' Responses to Item 1 (What do you Understand by Constructivism?)

All participants understood that knowledge is not transmitted passively but rather constructed actively through the participation of the students and the following are examples of some of their responses:

> As an epistemological paradigm, the central idea of constructivism is that individuals construct their own knowledge … it recognizes that the objective and value free nature of science is a myth popularized by positivism.

> Constructivism is a heterogeneous movement, which has many variants, such as: contextual, dialectical, empirical, Piagetian, pragmatic, radical, social, trivial, etc. The basic idea, however, is that an individual constructs her/his own knowledge.

> According to constructivism, learning is a social and dynamic process in which the learner constructs its significance.

These responses show that for most participants, constructivism had many forms and that students play an active part in the construction of their knowledge.

Note: In order to provide reliability of results (triangulation of data sources), participants' critical essays and interventions during the 17 class presentations were checked (see Multiple Data Sources, for a pool of 345 questions and comments). It was found that all 17 participants expressed similar ideas on at least two different occasions. It is important to note that Guba and Lincoln (1989) consider such "cross-checking specific data items of a factual nature" (p. 241) as an essential part of triangulation of data sources.

Participants' Responses to Item 2 (Why are there so many Forms of Constructivism?)

All participants understood that the different forms of constructivism in education are in response to attempts at integration of psychology of learning to the epistemology of the construction of knowledge. Just as there are epistemological differences with respect to the construction of knowledge, we inevitably have different forms of constructivism and following are some examples of participants' responses:

> It occurs to me that the different forms of constructivism are a consequence of the dynamics of progress in science; we are well aware of the ample range of considerations, conflicts and controversies in the last few decades. This provides evidence for the complex nature of scientific progress, and hence we are far from having theories that are definitive.

> Just as in science, the different postures in constructivism reflect the considerations and praxis of the scientist. In other words, science is not rigid and various methods can be used to solve the same problem.

> The different forms of constructivism can be understood as depicted by Hilary Putnam's "boat metaphor", in which people on different boats form part of a fleet and thus provide not only stimulus but also criticisms and sometimes people abandon one boat in favor of another. [Putnam's "boat metaphor" was used in Niaz et al. (2003), Reading 16, to illustrate critical appraisals of constructivism in science education. In both science and psychology, we may have the same data and still our interpretations may differ widely. The oil drop experiment (cf. Niaz, 2000a, 2005b) is a good example of how two leading scientists (Millikan and Ehrenhaft) interpreted very similar data in entirely different ways. Similarly, Piaget is critiqued primarily for his interpretations (cf. Carey, 1986). The "boat metaphor" thus helps to understand that parallel research programs are almost inevitable, especially due to the complexity of the problems.]

These responses show participants' understanding with respect to the conflicts and controversies involved in the construction of scientific knowledge and that this inevitably, leads to alternative forms of constructivism.

Note: In order to provide reliability of results (triangulation of data sources) participants' critical essays and interventions during the 17 class presentations were checked (see pool of 345 questions in Multiple Data Sources). It was found that all participants expressed similar ideas on at least two different occasions.

Participants' Responses to Item 3 (In your Opinion, can the Present State of Constructivism in Science Education be Considered as a Paradigm, as Conceived by Kuhn, 1962?)

As compared to Items 1 and 2, this item provided the participants with a bigger challenge and hence the need to understand constructivism within an epistemological perspective. Nine participants agreed (see Table 11.2) with the thesis that the present state of constructivism in science education can be construed as that of a Kuhnian paradigm and the following are examples of some of their responses:

> The presence of so many forms of constructivism (Piagetian, radical, social, etc.) demonstrates how each one of these [forms] critiques the other based on a particular paradigm. According to Kuhn's thesis of paradigmatic incommensurability this precisely shows that, scientific theories based on different paradigms cannot be compared objectively.

> Based on the idea that the degree of consensus within the scientific community denotes the paradigmatic character of a theory, the present state of constructivism in science education can be considered as a Kuhnian paradigm.

These responses contrast sharply with respect to those participants (n = 8) who disagreed with the thesis that constructivism in science education could be construed as a Kuhnian paradigm and the following are examples of some of their responses:

> Different forms of constructivism in science education represent alternative interpretations of progress in science based on limitations, conflicts and difficulties faced by the scientists. In contrast, Kuhn (1962) has emphasized the importance of "normal science" for science education, which ignores the rivalries and hence competing schools of thought.

> A paradigm represents a position that has acquired consensusswell this is what we have been doing so far, namely, teaching "normal science" and textbooks represent a good example of this consensus. However, the debate that started in the 1970s (Piaget, Ausubel) and continues with different forms of constructivism (radical, social, others), provides a scenario that can hardly be considered as that of consensus and hence lack of a paradigm.

Responses to this item show that participants were about equally divided with respect to depicting the present state of constructivism in science education as a

Kuhnian paradigm. Those in favor of this thesis argued that the many forms of constructivism was a manifestation of the incommensurability of scientific theories and hence could be understood as Kuhnian paradigms. Participants who disagreed with the thesis faced a dilemma: textbooks presented a consensus view that approximated Kuhn's "normal science", and still there were different forms of constructivism which reflected alternative interpretations of progress in science. The reference to textbooks and Kuhn's "normal science" is interesting as in a previous course various historical controversies and their presentation in textbooks were discussed (e.g., Niaz, 1998, 2000a).

Note: In order to provide reliability of results (triangulation of data sources) participants' critical essays and interventions during the 17 class presentations were checked (see Multiple Data Sources). It was found that both groups of participants (those who agreed or disagreed with the thesis) expressed similar ideas on at least two different occasions.

Table 11.2 Profile of Participants' Responses to Different Items (n = 17)

Participant	*Item 3*	*Item 4*	*Item 5*	*Item 6*	*Item 7*
1	Agreed	Integral	Critical	Agreed	Agreed
2	Agreed	Social	Uncritical	Disagreed	Agreed
3	Disagreed	Social	Critical	Disagreed	Agreed
4	Disagreed	Social	Uncritical	Disagreed	Agreed
5	Agreed	Integral	Critical	Disagreed	Agreed
6	Agreed	Integral	Critical	Disagreed	Agreed
7	Agreed	Social	Uncritical	Agreed	Agreed
8	Agreed	Integral	Critical	Disagreed	Agreed
9	Agreed	Integral	Uncritical	Disagreed	Agreed
10	Disagreed	Social	Critical	Agreed	Agreed
11	Agreed	Radical	Critical	Disagreed	Agreed
12	Disagreed	Social	Uncritical	Disagreed	Agreed
13	Disagreed	Ausubel-Piaget	Critical	Disagreed	Agreed
14	Disagreed	Social	Critical	Disagreed	Agreed
15	Disagreed	Novak	Critical	Disagreed	Agreed
16	Disagreed	Dialectic	Critical	Disagreed	Agreed
17	Agreed	Social	Critical	Disagreed	Agreed

Item 3: In your opinion, can the present state of constructivism in science education be considered as a paradigm, as conceived by Kuhn (1962)?
Item 4: Which form of constructivism do you prefer?
Item 5: Indicate one aspect of social constructivism that you do not share and suggest an alternative.
Item 6: Given the popularity of social constructivism, do you think it constitutes a paradigm (Kuhn, 1962)?
Item 7: If scientific knowledge is tentative, do you think that the present state of constructivism must also evolve toward other forms?

Notes
1. In Item 4, Integral means the integration of the different forms of constructivism.
2. In Item 5, critical means that the participants explicitly pointed out some aspect that they did not share. Uncritical means generally in agreement with social constructivism.

Participants' Responses to Item 4 (Which Form of Constructivism do you Prefer?)

This question was particularly difficult for the participants, as one may be critical of the different forms of constructivism and still have difficulty in having preference for a particular form. The following response from one of the participants illustrates the dilemma faced by most:

> Knowing the different forms of constructivism, it is difficult to have preference for any particular form. What I would prefer is the possibility of having available all the forms so that I can pick and choose different elements according to the dilemma faced in the classroom. I must add, however, that I am more inclined toward social constructivism. [Many participants expressed a similar concern and finally inclined toward some form of constructivism.]

Despite a critical appreciation of most forms of constructivism, all participants finally did express a preference for some form of constructivism: Social = 8, Integral = 5, Radical = 1, Novak = 1, Ausubel and Piaget = 1, Dialectic = 1. One of the participants who preferred social constructivism explained: "It provides an opportunity for interaction among students and teachers in a creative and dynamic process. The facet I like most is its emphasis on students' prior knowledge and experiences." The following is an example of a response from a participant who preferred integration of the different forms:

> Although the different forms of constructivism have their pro and contra, I would prefer a hybrid. Mixing of the different forms would adapt better to my classroom practice for the following reasons: classes can be heterogeneous, possibility of applying different methodologies and time available during the semester.

Note: In order to provide reliability of results (triangulation of data sources) participants' critical essays and interventions during the 17 class presentations were checked (see Multiple Data Sources). It was found that 15 participants expressed a preference for the same form of constructivism on at least four different occasions. Two of the participants expressed some ambiguity with respect to their preferred form of constructivism.

Participants' Responses to Item 5 (Indicate One Aspect of Social Constructivism that you do not Share and Suggest an Alternative)

As compared to Items 1–4 (Initial exam), this item formed part of the Final exam in the last session of the 11th week. Between the Initial and the Final exam, participants engaged in 36 hours of classroom presentations, discussions and question–answer sessions, which provided considerable amount of experience with respect to the issues being discussed and a good overview of the complexities involved. Interestingly, 12 participants critiqued social constructivism for having overemphasized the importance of prior knowledge of the students as the most important variable in the classroom. Of these 12 participants, 10 suggested that as an alternative it is necessary to take into consideration the cognitive development of the students (one participant suggested that the role of the teacher be emphasized). The following is an example of a response from this group of participants:

> Initially, I was more disposed towards social constructivism. However, as I have reflected, I consider that the cognitive development of a student is facilitated in the degree to which the social environment provides cognitive perturbations leading to: assimilation—accommodation—equilibration (Piaget).

Five participants were in general uncritical of social constructivism and only made minor observations for improvement and clarification, such as: two participants suggested that the rapidly changing nature of constructivism in science education led to heterogeneity and overlapping among different forms of constructivism and thus created confusion; one participant disagreed with the metaphor of "child as a scientist" and instead suggested "child as a developing scientist"; two participants considered that more work needs to be done in order to elucidate criteria for student evaluation. These participants referred to the dilemma faced by Martha (Tobin & LaMaster, 1995) and Sarah (Glasson & Lalik, 1993) that had been discussed in the methodology course in the previous year. Although both subscribed to social constructivism, Martha believed that the evaluation must include questions that only the very capable students could answer, whereas Sarah believed that students should be evaluated based on what they have learned.

Note: In order to provide reliability of results (triangulation of data sources) participants' critical essays and interventions during the 17 class presentations were checked (see Multiple Data Sources). The group of 12 participants who critiqued social constructivism for having overemphasized the prior knowledge of the students expressed similar ideas on at least two different occasions. The other group of five participants, who were generally uncritical of social constructivism, maintained a similar stance on at least two different occasions.

Participants' Responses to Item 6 (Given the Popularity of Social Constructivism, do you think it Constitutes a Paradigm, Kuhn, 1962?)

It is important to note that this item is quite similar to Item 3 (Initial exam). There are, however, two important differences: (a) in Item 3, participants were asked if they considered constructivism in general to be a Kuhnian paradigm, whereas in Item 6 the question refers to social constructivism; and (b) Item 6 formed part of the Final exam, which means participants had considerably more experience (36 hours of classroom presentations and discussions). Fourteen participants did not agree that social constructivism constitutes a paradigm and following are some examples:

> Social constructivism is very important, however, it cannot be considered a panacea, as in contraposition we also have other forms of constructivism. Observing from the Kuhnian point of view, social constructivism has not totally displaced another theory, but on the contrary has generated many controversies, which does not permit a consensus within the scientific community. [This response has important features: first, social constructivism has not displaced other forms of constructivism, as would be expected from the Kuhnian perspective, second instead of a consensus we have many controversies, third how the scientific community perceives the role of different forms of constructivism.]

> The Kuhnian position presupposes the idea of "normal science" which facilitates an emergent consensus leading to a paradigm. Social constructivism cannot be considered to be the consensus in science education. Furthermore, consensus at best is a transitory feature of scientific progress, both in science and education. [This response not only disagreed with respect to social constructivism being a Kuhnian paradigm, but rather goes beyond by suggesting a Lakatosian thesis, namely progress in science is not necessarily characterized by consensus but rather controversies.]

> There will always be paradigms as we are constantly engaged in controversies. We are always in the pre-paradigmatic stage, that is, engaged in the development of a research program. Thus the different forms of constructivism cannot be considered as paradigms but rather research programs that are constantly faced with controversies. [This response draws attention to the fact that we do not have to understand the present state of constructivism as necessarily a paradigmatic manifestation but rather a constant process of critical appraisals.]

Three participants agreed that social constructivism can be considered a paradigm in the Kuhnian sense and the following is an example:

> Observing how social constructivism is implemented in our educational system, especially at the secondary level, it can be considered as a paradigm. In my opinion, scientific progress consists of new paradigms that are more consistent and have more capacity for solving problems. In contrast, for many teachers social constructivism will solve all the problems in their classrooms. [This response raises an important issue for the teachers and classroom practice. Given the popularity of social constructivism, many teachers may construe this to be a "panacea" and thus consider its implementation as simple and straight forward. It is interesting to note that this participant (#10, see Table 11.2) had disagreed on Item 3, as at that stage constructivism

did not seem to be very popular. This precisely shows the importance of understanding scientific progress in the historical context, that is the tentative nature of science and hence the importance of critical appraisals both in science and education. See Table 11.1 for details.]

At this stage it is important to compare results obtained in Items 3 and 6, as in both items participants were asked to evaluate the present state of constructivism in science education as a Kuhnian paradigm. On Item 3 (Initial exam), nine participants agreed that constructivism could be conceived of as a paradigm and eight disagreed. In contrast, on Item 6 (Final exam), only three participants agreed that constructivism could be conceived as a paradigm and 14 disagreed (see Table 11.2). Given the importance for science educators to understand the philosophical underpinnings of progress in science, this could be interpreted as a progressive conceptual change, providing greater understanding of constructivism as a scientific theory. Failure to understand that research in science (and education) is rarely free of controversy, and that progress is intricately woven with the confrontation of ideas, can lead the teachers to consider constructivism as a plethora of "truths" that can be memorized and applied as an algorithm. Recent literature in philosophy of science has recognized the role of controversies in very clear terms:

> Controversies are fated to arise when the new claim does not fit with the accepted endoxa and the community cannot neglect it. This creates a problem of consistency that cannot be solved unless the set of accepted endoxa is altered in some way or another. Rival parties then emerge with different solutions as to how a remedy can be found for the anomaly, and which endoxa need to be altered or modified. The deeper and wider the scope and effects of the needed alterations, the greater and more important the controversy.
>
> (Machamer, Pera & Baltas, 2000, p. 16)

Although these philosophers of science were representing the role of controversy in scientific progress, it reflects quite cogently the development of constructivism in science education during the last three decades.

Note: In order to provide reliability of results (triangulation of data sources) participants' critical essays and interventions during the 17 class presentations were checked (see Multiple Data Sources). It was found that 14 participants who disagreed also expressed similar ideas on at least two different occasions. Again, the three participants who agreed also expressed similar ideas on two different occasions.

Participants' Responses to Item 7 (If Scientific Knowledge is Tentative, do you think that the Present State of Constructivism must also Evolve Towards Other Forms?)

Interestingly, all 17 participants agreed that the present form of constructivism must eventually develop and progress toward other forms of constructivism and the following are examples of some of their responses:

> Due to the tentative nature of scientific knowledge—when constructivism does not provide convincing answers for teaching science—when new evidence contrasts with that provided by constructivism—at that stage constructivism will stop being a panacea and undoubtedly must transform into other forms. [The degree to which this response reflects the same basic idea as that presented by philosophers of science, Machamer et al., 2000, cited above, is thought-provoking indeed, as these teachers were not aware of this publication.]

> The multiple contradictions and controversies that surround constructivism, show that there is no absolute truth and if scientific knowledge is tentative, then this provides the base for the advance of constructivism. [It is important to note that participants in this course being chemistry teachers constantly referred to the historical episodes that are represented in Table 11.1 as markers (both for science and science education). All participants had a fair grasp of the historical context in which these atomic models developed through readings such as: Lin & Chen (2002), Niaz (1998, 2000a). For example, when Rutherford presented his model of the atom, did that mean that the previous Thomson model was not correct or not true? Participants were fully aware that in this particular context we cannot expect science to provide an "absolute truth".]

> The present stage of constructivism must change and improve—this is the nature of science and that of the different epistemological, psychological and philosophical currents. Now, where do we go? I do not know! [This response, like those of other participants, shows that these teachers have grasped the idea that if constructivism is a scientific theory then it must continually change towards more progressive forms. Nevertheless, this particular participant ends on an uncertain note ("I do not know") as it is precisely the teachers and classroom practice that can play a crucial role in the future development of constructivism.]

Note: In order to provide reliability of results (triangulation of data sources) participants' critical essays and interventions during the 17 class presentations were checked (see Multiple Data Sources). It was found seven participants had expressed similar views on at least four different occasions (this group of participants had disagreed on Items 3 and 6). The remaining 10 participants expressed similar views on at least two different occasions.

Evidence for Greater Understanding of Constructivism

Research design of this study provided evidence for transitions leading to greater understanding of constructivism by the participants. Different items of the Initial exam (first day of the last week) and the Final exam (last day) were carefully selected for this purpose. As the Initial and the Final exams were separated by 36 hours of classroom activities and discussions, it is plausible to suggest that the following comparisons provide evidence for greater understanding:

a. *Comparison of Item 3 (Initial exam) and Item 6 (Final exam)*: In both items participants were asked to evaluate the present state of constructivism as a Kuhnian paradigm. On Item 3, nine participants agreed that constructivism could be conceived as a Kuhnian paradigm. In contrast, on Item 6, only three participants agreed that social constructivism could be conceived as a paradigm (see Table 11.2). Most participants who experienced a change reasoned cogently with respect to controversies in the development of constructivism and hence it could not be construed as a paradigm. This is a clear evidence for greater conceptual understanding;
b. *Comparison of Item 4 (Initial exam) and Item 5 (Final exam)*: On Item 4, eight participants selected social constructivism as their preferred form of constructivism. On Item 5, based on the experience gained in classroom discussions, four of these eight participants were critical of social constructivism. This once again showed greater understanding of the issues involved;
c. *Comparison of Item 2 (Initial exam) and Item 7 (Final exam)*: On Item 2, all participants understood that the different forms of constructivism were a consequence of the difficulties/contradictions involved in understanding scientific change and hence a proliferation of theories in both science and education. On Item 7, all participants went beyond by understanding that the contradictions faced by the existing forms of constructivism would inevitably lead to new forms of constructivism.

Conclusions and Educational Implications

The course content and the seven evaluation items used in this study served as probes and at the same time provided stimulus to participant teachers to think beyond their initial understanding of constructivism. Item 1 facilitated the understanding that construction of knowledge requires active participation of learners. Item 2 suggested that the different forms of constructivism represent competing and conflicting interpretations of progress in science. On Item 3, participants were about equally divided with respect to accepting the present state of constructivism in science education as a Kuhnian paradigm. Given the popularity and influence of Kuhnian philosophy in education (Loving & Cobern, 2000; Matthews, 2004), the position of those who agreed with the thesis is not surprising. Interestingly, those who disagreed argued cogently with respect to chemistry textbooks being representative of Kuhn's "normal science" and thus ignoring alternative interpretations of progress in science. Item 4 referred to participants' preference for a particular form of constructivism, and eight preferred social constructivism, whereas five suggested integration of the different forms according to the needs and requirements of classroom practice. Items 1–4 formed part of the Initial exam.

Item 5 formed part of the Final exam (which provided more time and opportunities to think and reflect) and asked participants to indicate one aspect of social constructivism that they did not share. Twelve teachers critiqued social constructivism for overemphasizing the prior knowledge of the students and suggested that cognitive development needs to be taken into consideration. Interestingly, four of these 12 teachers had selected social constructivism as their preference in Item 4.

Item 6 explicitly asked participant teachers if they considered social constructivism as a Kuhnian paradigm. Fourteen participants disagreed (three agreed with the thesis) and argued cogently by pointing out that: (a) social constructivism, despite its popularity, has not displaced other forms of constructivism; (b) there is no consensus in the science education community with respect to constructivism; and (c) we are always engaged in controversies based on different research programs. Interestingly, of the 14 participants who disagreed with the thesis that social constructivism constituted a Kuhnian paradigm, seven had previously (Item 3) agreed that constructivism constituted a paradigm. This clearly facilitated greater understanding of the teachers with respect to constructivism and progress in science. Interestingly, a recent critique has analyzed constructivism in science education as a Lakatosian research program and suggested that it may be "degenerating" or on the verge of "a more promising candidate" (Taber, 2006, p. 209).

Item 7 clearly provided teachers the experience necessary to understand that if scientific knowledge is tentative, then constructivism must also evolve toward more progressive forms. Arguments presented by some of the participants provide evidence for such a conceptual change, such as: (a) when new evidence contrasts with that provided by the existing form of constructivism; (b) constructivism cannot be a panacea; (c) contradictions faced by constructivism provide the base for its advance; (d) where do we go from here? This looks more like an outline of a new research program and possibly a response to, whither constructivism? The most important aspect of these responses by the teachers is the understanding that it is precisely the contradictions faced by the existing forms of constructivism that would facilitate alternative and rival theories, leading to new advances. This may require more eclectic and perhaps pragmatic approaches, which has been cogently argued by Perkins (1999):

> The term *constructivism*, with its ideological overtones, suggests a single philosophy and a uniquely potent method—like one of those miracle knives advertised on late-night TV that will cut anything, even tin cans. But we could look at constructivism in another way, more like a Swiss army knife with various blades for various needs ... it's high time we got pragmatic about constructivism.
>
> (p. 11)

In other words, if a particular approach does not solve the problem, try another—more structured, less structured, more discovery-oriented, less discovery-oriented, whatever works. And when knowledge is not particularly troublesome for the learners in question, perhaps it would be better to forget about constructivism and consequently teaching by telling may be the alternative (cf. Perkins, 2006).

This study has implications for teacher education and in-service training. It is important for teachers to understand that constructivism cannot be applied as an algorithm in the classroom, leading to a panacea (cf. Tobias & Duffy, 2009). On the contrary, based on an exploration of the nature of science and its critical appraisal (cf. Units 1 and 2 of Course Content), this study facilitated a critical appraisal of the current status of constructivism in science education (cf. Unit 3 of Course Content). For example, if the teachers consider social constructivism as a Kuhnian paradigm, that may lead them to ignore the controversies with respect to the different forms of constructivism in science education. Even Glasersfeld has recently acknowledged that, "radical constructivism is not a dogma and it does not claim to be 'true'" (interview with Cardellini, 2006, p. 185).

Teachers in this study were exposed to a wide range of opinions with respect to constructivism, such as: Abd-El-Khalick et al. (2004), Gil-Pérez et al. (2002), Marín et al. (1999), Martínez (1999), Matthews (2004), Moreno and Waldegg (1998), Niaz et al. (2003) and Osborne et al. (2003). It is important to note that teachers made considerable effort to understand these readings based on formal presentations (PowerPoint), followed by question–answer sessions and critical essays. Teachers found these readings interesting as they could clearly identify the instructor's position on various issues and contrast them with other authors. This encouraged the participants to be critical as they found multiple opinions about the same issue. At this stage it is necessary to clarify that this study does not espouse a philosophically relativist theory. On the contrary, both science and cognitive psychology provide many good and useful answers to the problems we face. Science curricula and textbooks in most parts of the world (Niaz, 2008a; 2009a) bear witness to the fact that students and teachers do not understand the dynamics of scientific progress, either in science or in psychology/education. In other words, most educational systems are simply prescriptive, that is, Bohr's atomic model was wrong, Piaget's theory of cognitive development and especially views on constructivism were wrong, etc. It would be more helpful if we make students and teachers understand as to how atomic models have evolved since Bohr and similarly how Piaget's theory must be complemented with more recent developments in cognitive psychology (progressive problemshifts, Lakatos, 1970).

Finally, a word of caution is necessary as understanding constructivism inevitably involves nature of science and psychology of learning, which in turn requires multiple perspectives. In this particular study, constructivism and its critical appraisal were developed within a history and philosophy of science perspective and hence the impression that only nature of science helps understand constructivism. In other words, our current models are partial and incomplete (tentative nature of scientific knowledge), and further research and critical appraisals are needed to refine our models toward a more complex all-embracing and progressive constructivist approach. Although teachers in this study had a fair amount of exposure to these areas of knowledge in this and the previous courses, it is plausible to suggest that such courses and readings become a permanent part of in-service teacher-training programs.

Chapter 12

Conclusion

Methodologists Need to Catch Up with Practicing Researchers

An Overview

A classroom teacher is generally faced with numerous problems. The traditional approach is to provide the teacher with textbooks, worksheets, lab manuals and finally the lecture with the following guideline: this is what science has found and so you must learn and teach. An alternative approach to solve these problems would be providing the teachers an opportunity to do research based on the understanding that science is tentative, where different interpretations are offered and views held despite seeming refutation. This would provide the students and teachers an understanding of the dynamics of scientific progress (Niaz, 2009a) and a glimpse of "science in the making". Teacher as a practicing researcher thus could provide the methodologists an insight with respect to what happens in the classroom. Consequently, researchers in teacher education first need to ask important questions that need answers and then come up with a combination of research genres that are appropriate (Borko et al., 2007). This is all the more important if we recognize the pitfalls associated with the traditional positivist view of the scientific enterprise, which clearly shows the need to go beyond and accept perspectives that align with postpositivism (Phillips & Burbules, 2000). In this context, Giere (2006) provides sound advice:

> I wish to reject objective realism but still maintain a kind of realism, a perspectival realism, which I think better characterizes realism in science. For a perspectival realist, the strongest claims a scientist can legitimately make are of a qualified, conditional form: "According to this highly confirmed theory (or reliable instrument), the world seems to be roughly such and such." There is no way legitimately to take the further objectivist step and declare unconditionally: "This theory (or instrument) provides us with a complete and literally correct picture of the world itself."
>
> (pp. 5–6)

At this stage it would be helpful to sit back, think retrospectively and consider the following: how many teachers around the world present students with "a complete and literally correct picture of the world"? No wonder Tobias (1993) refers to this approach as the "tyranny of technique" which deprives students from the profound intellectual experience they had expected from studying science. Science if studied within this perspective can provide not only a profound and exciting intellectual challenge but also a means to emancipation (liberation from received wisdom, Collins, 2000). Interestingly, according to Phillips (2005b) the traditional approach to research (devoid of a historical perspective) would not consider Galileo, Harvey, Newton, Pasteur and Darwin as rigorous researchers.

This book is based on the following epistemological guidelines: (a) it is the problem to be researched that determines the methodology to be used; (b) a historical reconstruction of a scientific theory can determine the different sources that contributed to its development; and (c) discussion of the historical episodes based on interactions among classroom teachers can facilitate the elaboration of new teaching strategies. These guidelines have been followed in this book while discussing the different historical episodes, which have important implications for teacher training and going beyond the traditional teaching approach.

It is important to note that given the opportunity, teachers do understand that in order to fire the imagination of the students we need to go beyond "normal science" (Kuhn, 1962) and provide them with different scenarios of science-in-the-making (Chapter 2). Based on the arguments presented in this book, it is plausible to construct the following sequence: Kuhn's "normal science" manifests itself in the science curriculum and textbooks through the scientific method—teaching normal science leads to memorization of science content with little understanding—eliminating normal science would facilitate the inclusion of controversies and thus provide students a glimpse of what science is all about. Consequently, if we want our students and teachers to scrutinize and understand scientific practice, then a revision of the science curriculum is necessary.

A major argument in favor of mixed methods (integrative) research programs is that it provides a rationale for hypotheses, theories and guiding assumptions to compete and provide alternatives. It is concluded that such research programs (not paradigms) in education can facilitate the construction of robust strategies, provided we let the problem situation (as studied by practicing researchers) decide the methodology (Chapter 3).

Based on the results obtained from Chapter 4, it is concluded: (a) participants were able to understand the basic ideas of constructivist philosophy and its pedagogical implications; (b) role of behavioral objectives in actual educational practice was questioned; (c) integration of qualitative and quantitative research methods was considered to be an alternative to the current debate about the replacement of one method by the other; (d) participants considered the dilemma of evaluating students based on what they have learned or what they should have learned, within the social constructivist framework and generally favored the former; (e) most of the participants were reluctant to accept constructivism as a form of positivism, a controversial thesis that has some support in the research literature. Given the importance of alternative approaches to growth and meaning of knowledge, it is important that teachers be aware of conflicting situations in the classroom that refer to: objectivity, scientific method, qualitative-quantitative methods, relationship between method and problem, evaluation, and a critical appreciation of constructivism.

Given the importance of qualitative research, an important issue of concern is the degree to which its findings can be generalized (Chapter 5). Based on the results obtained it is concluded: (1) almost 91% of the teachers agreed that external generalization in a different social context is feasible; (2) almost 63% of the participants used a fairly inconsistent approach, that is, in a theoretical context agreed that qualitative research cannot be generalized and still when asked with respect to the experience of two particular teachers, agreed that generalization was possible; and (3) almost 28% of the participants used a consistent approach. This clearly shows the difficulties teachers have with respect to how and when some degree of generalization is not only possible but also desirable. Some of the reasons provided by the participants as to why generalization was feasible are discussed. An analogy is drawn with respect to Piaget's methodology (basically qualitative research), namely, it was not based on random samples or statistical treatments and still his *oeuvre* has been generalized (criticisms not withstanding) in both the psychology and educational literature.

Despite its popularity there are misunderstandings with respect to various aspects of qualitative research methodology in education, which were the subject of Chapter 6. Based on the results obtained it is concluded: (a) most participants understood that the problem to be investigated precedes the method and determines the methodology to be used; (b) as all observations are theory-laden, it is preferable that interpretations based on both qualitative and quantitative data be allowed to compete in order to provide validity to our research findings; (c) the difference between validity and authenticity was controversial and most participants considered the need for interpreting data and hence favored authenticity; (d) discussions led to the idea of "degrees of validity" as both validity in the quantitative sense and authenticity in the qualitative sense ultimately depend on critical appraisals of the community; (e) generalizability of results obtained from qualitative studies was a controversial topic and most participants agreed that it is not desirable to generalize; (f) discussions suggested an alternative: in both qualitative and quantitative research generalizability is possible, provided we are willing to grant that our conceptions/theories are not entirely grounded in empirical evidence but rather on the degree to which the community of researchers can uphold such a consensus; (g) most teachers considered the use of participant observation in qualitative research as non-controversial. Class discussions led to the understanding that emphasizing observations may lead us to the Aristotelian ideal of empirical science; (h) formulation of hypotheses, manipulation of variables and the quest for causal variables were considered by many teachers to be equivalent to the scientific method. Discussions facilitated the understanding that this led to idealization (a methodology used by both the natural and social scientists) and thus helped to reduce the complexity of a problem.

Ability to elaborate and differentiate between a *hypothesis* and *prediction* is an important part of the research methodology in educational research (Chapter 7). Results based on written responses showed that most teachers (~60%) do not understand the difference between a hypothesis and a prediction. It was also observed that many teachers did provide a satisfactory description of what they considered to be a hypothesis and a prediction. However, the difficulty for the teachers consisted in operationalizing (elaborating and understanding) the difference between a hypothesis and a prediction. Some teachers explicitly elaborated and classified a prediction as a hypothesis and at times the teachers elaborated the two in the same manner, without being aware of the contradiction. This study has educational implications by showing that, just like students, teachers also have difficulties with the elaboration and understanding of a hypothesis and a prediction. Given the importance of such concepts for all research programs it is essential that appropriate teaching strategies be implemented. Results from the follow-up training study, based on written responses, showed that the experimental group performed better than the control group in elaborating both a *hypothesis* and *prediction*. The difference was, however, statistically significant only for *prediction*. Despite improvement, almost 50% of the teachers still had difficulty in formulating a *hypothesis* and 40% in formulating a *prediction*.

Alternative interpretations of data that facilitate conceptual understanding are an important aspect of various historical episodes. Most of the research dealing with conceptual change among students has, however, ignored this important facet of scientific practice. The study in Chapter 8 was designed to familiarize participating teachers with this important aspect and some of the educational implications are: (a) similar to a scientist a student can "live" with two rival theories simultaneously and as the student enriches his cognitive repertoire the conflict can perhaps be resolved; (b) resolution of a conflict may not follow a logical pattern of reasoning but rather a slow process (based on motivational, intuitive and affective factors) in which the "hard-core" (negative heuristic, Lakatos, 1970) of beliefs slowly "crumbles"; (c) in science there is no absolute truth, nor a "scientific method" and consequently there cannot be rules, methods, algorithms or predetermined steps for introducing conceptual change; (d) teachers' epistemological outlook is crucial in order to facilitate conceptual change.

The basic premise of the study in Chapter 9 was that, given the opportunity to reflect, discuss and participate in a series of course activities based on various historical controversies, teachers' understanding of nature of science can be enhanced. Based on the results obtained it is suggested that this study facilitated the following progressive transitions in teachers' understanding of nature of science: (a) problematic nature of the scientific method, objectivity and the empirical basis of science; (b) myths associated with respect to the nature of science and teaching chemistry; (c) science does not develop by appealing to objectivity in an absolute sense, as creativity and presuppositions also play a crucial role; (d) the role of speculation and controversy in the construction of knowledge based on episodes from the science curriculum; (e) how did Bohr confirm his postulates? This goes beyond the treatment in most textbooks and provides an opportunity to ground educational practice in an historical episode; (f) differentiation between the idealized scientific law and the observations.

There has been controversy in the science education literature with respect to what "ideas-about-science" should be taught in school science and the objective of Chapter 10 was to familiarize teachers with this controversy and thus facilitate a better understanding of the underlying issues. Based on the results obtained this study suggested the following educational implications: (a) experimental data need to be interpreted carefully due to under-determination of theories by data; (b) Kuhn's normal science manifests itself in the science curriculum through the scientific method and wields considerable influence; (c) trilemma posed by Collins (2000), namely, creation of new knowledge ⇔ Kuhn's normal science ⇔ teaching nature of science, provided a big challenge and was thought-provoking; (d) of the different aspects of nature of science suggested by experts, these teachers endorsed the following as most important: creativity, historical development of scientific knowledge, diversity of scientific thinking and scientific method and critical testing; (e) with respect to the contradiction between the positions of Lederman et al. (2002) and Osborne et al. (2003), few supported the position of latter, namely, inclusion of scientific method in the classroom and a majority supported the former, namely, scientific method as a myth; (f) participants were critical of the present stage of research with respect to the scientific method and suggested the introduction of history, philosophy and epistemology of science to counteract its influence.

Constructivism has been the subject of considerable controversy in the science education literature and Chapter 11 provided the teachers an opportunity to consider this as a manifestation of the tentative nature of scientific knowledge. Based on results obtained it is plausible to suggest that participant teachers experienced the following transitions leading to greater understanding, as they acquired experience with respect to constructivism: (a) active participation of students as a pre-requisite for change; (b) different forms of constructivism represent competing and conflicting interpretations of progress in science; (c) acceptance of the present state of constructivism as a Kuhnian paradigm; (d) social constructivism as the preferred form of constructivism; (e) critical appraisal of social constructivism; (f) despite its popularity social constructivism does not constitute a Kuhnian paradigm (due to controversies there is no consensus in the science education community); (g) contradictions faced by constructivism in science education provide the base for its advance and evolution towards more progressive forms, and hence the need to consider, whither constructivism?

Normal Science versus Science in the Making

There is a fundamental tension between the problematic nature of Kuhn's "normal science" and instead emphasizing "science in the making", namely the "how" and "why" of scientific progress, based on a Lakatosian perspective (Chapter 2). The Kuhnian perspective in general emphasizes the right answer (e.g., current atomic model) based on current ideas or theories with respect to a particular topic of the curriculum. Mastery of the right answer avoids alternative approaches and inevitably leads to the elaboration and memorization of algorithms. Furthermore, Kuhn's incommensurability thesis presents a vision of progress in which one science can accommodate only one paradigm. A teacher following the Kuhnian perspective thus may use the textbook, worksheet, lecture or other implements to present what science has found to be the case, and not how scientists constructed an idea, hypothesis or theory. On the contrary, the Lakatosian perspective would emphasize conflicting frameworks based on rival hypotheses that inevitably lead to alternative interpretations and manifest the tentative nature of science, that is "science in the making". A teacher following this perspective would be more conducive to facilitate debate between opposing ideas that may be expressed by different groups of students which may facilitate the generation of new perspectives and ideas. This understanding of "science in the making" based on the tentative nature of science is critical for teacher training and some of the examples are provided in this section.

Understanding alternative interpretations of conceptual change was the subject of Chapter 8. This provided classroom teachers an opportunity to evaluate the different strategies for conceptual change. In one of the research questions participating teachers were asked to consider the following dilemma and suggest alternatives: If a student thinks that scientific knowledge consists of facts, formulae and data then his understanding of science would be different from that of a student who has more adequate epistemological conception, based on the dynamics of scientific progress (Niaz, 2009a). Interestingly, the course provided the teachers the background knowledge to argue that the emphasis on facts and formulae reflected the conductivist legacy and hence the need to go beyond. Furthermore, these teachers considered that a more adequate epistemological conception would deal with the question: how did we come to have a certain piece of knowledge? In order to respond to this question, teachers suggested a historical reconstruction (based on controversies and rival hypotheses) to facilitate greater conceptual understanding. An important and novel feature of this response is the possibility of an alternative that can compete with the traditional teaching strategy based on facts and formulae. This Lakatosian perspective goes beyond the conceptual change model of Posner et al. (1982) and its revised version (Strike & Posner, 1992), both of which are heavily influenced by Kuhnian philosophy of science.

Understanding progressive transitions in teachers' understanding of nature of science was the subject of Chapter 9. In one of the research questions (2a) participating teachers were asked if they considered chemistry textbooks to present science as an illustration of the scientific method (a well defined stepwise procedure). One of the teachers responded by citing Bohr's postulates as an example of how textbooks present a scientific theory. It is important to note that the teacher went on to present the following dilemma: a teacher could wonder as to how Bohr confirmed his postulates. At this stage it would be interesting to recall Bohr's (1913) original work, presented in most science textbooks as the following postulates: (1) the electron in an atom has only certain definite stationary states (fixed energy) of motion allowed to it; (2) when an atom is in one of these states it does not radiate energy. When an electron in an atom changes from a high-energy state to a state of lower energy, the atom emits a quantum of radiation whose energy (hv) is equal to the difference in the energy of the two states; (3) in any of these stationary states the electron moves in a circular orbit about the nucleus; and (4) the states of allowed electronic motion are those in which the angular momentum of the electron is an integral multiple of $h/2\pi$. Most science textbooks, however, make no mention of the origin of these postulates, much less offer an explanation (cf. Niaz, 1998; Rodríguez & Niaz, 2004a). It is plausible to suggest that the dilemma posed by the teacher in this study is one way of going beyond the traditional "scientific method" and understanding "science in the making". Most science textbooks consider the scientific method to be a recipe-like procedure, which allows scientists to elaborate theories and laws infallibly. The teacher's response in this case is all the more significant for the following reason: The research question itself does not refer to Bohr's postulates. It was the teacher's ingenuity which facilitated an understanding of progress in science not as a recipe-like procedure, but by posing a dilemma: what was the warrant which allowed Bohr to enunciate his postulates? History of science shows that Bohr did not have the necessary experimental evidence to formulate his postulates, and it is precisely this part of Bohr's work that Lakatos (1970) considers the "negative heuristic" (i.e., imperative of presuppositions), which cannot be refuted even if it lacks experimental confirmation (i.e., quantitative imperative). The tension between the quantitative imperative and the imperative of presuppositions is an important driving force in scientific progress (Niaz, 2005a). Most science textbooks and even curricula simply require students to memorize Bohr's famous postulates as algorithms. On the contrary, it is plausible to suggest that a discussion of Bohr's four postulates based on an HPS perspective can provide teachers an opportunity to facilitate a better understanding of how a scientist goes about solving complex problems in the absence of convincing experimental evidence. This clearly shows the difference between Kuhn's normal science (learning Bohr's postulates) and "science in the making", that is, what was the warrant for Bohr's postulates. A modern commentator has provided a historical perspective with respect to Bohr's famous postulates in cogent terms:

> The first assumption [postulate] is the existence of stationary states, the second is the frequency rule. Bohr regarded them as the unshakable pillars of his theory. They were indeed more directly related to experiments than other assumptions of his theory. Until at least 1925, they remained the two basic postulates of the quantum theory, despite the vicissitudes of most other assumptions.
>
> (Darrigol, 2009, p. 154)

This discussion shows that most educational systems and science curricula are prescriptive, that is Bohr's atomic model was wrong, Piaget's theory of cognitive development and especially views on constructivism were wrong, and so on. It would be more helpful if we make students and teachers understand as to how atomic models have evolved since Bohr, and similarly how Piaget's theory must be complemented with more recent developments in cognitive psychology (e.g., Pascual-Leone's 1978 theory of cognitive development). This increase in heuristic or explanatory power of a theory is particularly easy to understand as a progressive problemshift (Lakatos, 1970). A recent study based on a historical reconstruction has illustrated how in the case of the atomic models it would be helpful to introduce the Bohr–Sommerfeld model, as an attempt to overcome the difficulties associated with the Bohr model and thus go beyond (Niaz & Cardellini, 2010).

Difficulties involved in introducing nature of science in the classroom, was the subject of Chapter 10. On the one hand, school science generally tries to inculcate a dogmatic and authoritarian approach (this is known to be true and so you must learn), whereas science also presents a popular culture that promotes emancipation based on scientific discoveries. The first alternative generally espouses the scientific method, approximates Kuhn's "normal science" and hence problematic. In one of the research questions (4) participating teachers were asked to resolve the dilemma with respect to the importance of the scientific method in the science curriculum. One group of science educators (Osborne et al., 2003) considers the "scientific method" to represent the "central thrust of scientific research". In contrast, Lederman et al. (2002) consider a "recipe-like stepwise procedure" a myth. Teachers were asked to resolve this dilemma between the two groups of science educators. Interestingly, one group of teachers tried to reconcile the two positions by suggesting that scientists do follow certain procedures. However, a stage comes when the procedures break down and a controversy ensues. One teacher suggested, "If there had been an infallible scientific method, both Thomson and Rutherford could have met over dinner and resolved the controversy" (this refers to atomic models, cf. Niaz, 1998). Another group of teachers went beyond by supporting the position of Lederman et al. (2002) and at the same time suggesting that this was an important issue and hence it was necessary to achieve consensus in the science education community. This group of teachers cited various episodes from the history of science to show how the scientific method did not work. Windschitl (2004) has presented this dilemma in succinct terms and perhaps can provide grounds for consensus among science educators:

> The Scientific Method (making observations, developing a question, constructing hypotheses, experimenting, analyzing data, drawing conclusions) is often portrayed in textbooks as a linear procedure; however, this characterization and even the label itself are misrepresentations. The process of hypothesis testing in science is not a linear one in which each step is a discrete event whose parameters are considered only after the previous step is complete. In authentic scientific practice, multiple steps or phases are often considered in relation to one another at the outset of the investigation.
>
> (p. 483)

Authentic scientific practice referred to by Windschitl (2004), approximates to what Niaz (2010) has recently suggested as "teaching science as practiced by scientists", which would be more motivating for students and thus facilitate a better understanding of progress in science.

Constructivism has been the subject of considerable controversy in the science education literature (Niaz et al., 2003; Tobias & Duffy, 2009). As a scientific theory constructivism reflects the tentative nature of scientific knowledge and this was the subject of Chapter 11. This study draws an explicit parallel between scientific progress, which is characterized by its tentative nature (based on conflicts, rivalries and alternative interpretations of data, cf. Niaz, 2009a) and the different forms of constructivism. In one of the research questions (6), participating teachers were asked if they considered social constructivism as a Kuhnian paradigm. A majority of the participants disagreed with this proposition and provided cogent reasons: (a) social constructivism, despite its popularity has not displaced other forms of constructivism; (b) there is no consensus in the science education community with respect to constructivism; and (c) we are always engaged in controversies based on different research programs. This clearly shows that participating teachers progressed toward an understanding of nature science in which consensus is not necessarily an end in itself. On the contrary, both scientists and researchers in science education continue to argue and critique the existing theories and thus facilitate the understanding that progress in science is tentative. This clearly shows that "science in the making" can provide students and teachers a better alternative than regurgitating experimental details of known facts, theories and laws based on Kuhn's "normal science".

Teaching Science as Practiced by Scientists

Peter Medawar, a leading scientist (Nobel Laureate in medicine, 1960) in his *The art of the soluble*, has asked a very pertinent question: what kind of act of reasoning leads to scientific discovery? Medawar (1967) himself responded to this question in terms that may surprise many scientists, science teachers and even curriculum developers:

> It is no use looking to scientific "papers," for they not merely conceal but actively misrepresent the reasoning that goes into the work they describe. If scientific papers are to be accepted for publication, they must be written in the inductive style. The spirit of John Stuart Mill glares out of the eyes of every editor of a Learned Journal ... Only unstudied evidence will do—and that means listening at a keyhole.
>
> (p. 151)

This raises important and critical issues: do scientific "papers" misrepresent the very process that they undertake to understand and how many scientists would tolerate an intruding historian, philosopher of science or science educator to listen at the keyhole? This echoes Phillips' (2005a) call to educators to follow the lead of philosophers of science (Popper, Kuhn, Lakatos, Holton, Cartwright and Galison), who started to change the philosophical landscape about six decades ago (cf. Chapter 3). To respond to Medawar's critique, Holton (1978b) has suggested that besides the "papers", scholars can study:

> letters, autobiographical reports cross-checked by other documents, oral-history interviews conducted by trained historians, transcripts of conversations that took place in the heat of battle at scientific meetings, and, above all, laboratory notebooks—firsthand documents directly rooted in the act of doing science, with all the smudges, thumbprints, and bloodstains of the personal struggle of ideas.
>
> (p. 25)

Indeed, this is a challenging and impressive agenda. Holton (1978a, 1978b, chapter 2) himself provided the lead by reconstructing the controversy between Robert Millikan and Felix Ehrenhaft, with respect to the determination of the elementary electrical charge, by consulting Millikan's hand-written notebooks at CALTECH (for details, cf. Niaz, 2005b). Without being exhaustive, other examples of such historical approaches are provided by the work of Galison (1987) and, more recently, Collins (2004).

Based on Phillips' (2005a) suggestion, one way to change the educational landscape is to familiarize teachers with developments in history and philosophy of science and develop a framework for classroom instruction (Chapters 2 and 3). The main premise of this book is to first highlight the problems and then to use the framework for developing possible alternatives based on classroom discussions. Teachers played a crucial role in understanding the problems within the context of the framework by participating actively in various activities. Participants were not only exposed to the controversies but also encouraged to support, defend or critique the different interpretations. Just like the historical reconstructions discussed in class provide a glimpse of "science in the making", all chapters of this book facilitate an understanding of how teachers interact to critically appraise dynamics of scientific progress. In subsequent chapters this framework has been used to facilitate a better understanding of problems faced by the teachers in the classroom.

Chapter 4 provided the teachers an opportunity to understand that given the importance of alternative approaches to growth and meaning of knowledge, it is important that teachers be aware of conflicting situations in the classroom that refer to "objectivity", "scientific method", qualitative-quantitative methods, relationship between method and problem, evaluation, and a critical appreciation of constructivism.

A major argument of qualitative researchers for not generalizing from qualitative studies is that this research is not based on sufficiently representative samples and adequate statistical controls (Chapter 5). In this context, it is pertinent to ask whether most of Piaget's work was based on representative samples? A review of the literature shows that this was not the case. So how did Piaget's work come to be generalized and accepted by the educational research community? A plausible reason is provided by Piaget's differentiation between the epistemic and the psychological subjects and that his methodology represented an example of Galilean idealization (discussed in Chapter 2). Epistemic subject refers to the underlying rationality (universal scientific reason) ideally present in all human beings. In other words, Piaget was not studying the average of all human abilities (hence lack of statistical treatments and random samples), but rather the ideal conditions under which a psychological subject (a particular person) could perhaps attain the competence exemplified by the epistemic subject. Teaching science as practiced by scientists, thus opens the possibility of generalization from qualitative studies in education.

Observations, in both qualitative and quantitative educational research are theory-laden and hence subject to scrutiny by the scientific community (Chapter 6). Data in themselves do not constitute "science", rather it is the arguments (interpretations) researchers put forward to convince their peers that help to construct science. Participant observation, an important research tool used by qualitative researchers is open to critical appraisal as these observations are also subject to the presuppositions of the researcher. Based on these considerations, it is the problem to be researched that determines the method in educational research.

The role played by alternative interpretations of data is an important part of scientific progress. Chapter 8 provided participating teachers an opportunity to scrutinize the historical record based on different interpretations of the same data in various historical episodes, such as: Thomson's hypothesis of compound scattering and Rutherford's hypothesis of single scattering to explain alpha particle experiments; Millikan's hypothesis of a universal electrical charge (electron) and Ehrenhaft's hypothesis of fractional charges (sub-electron) to explain the oil/metal drop experiments; Caloric and kinetic theories put forward to understand the same heat phenomena. There is a common thread running through these historical episodes, namely resolution of conflict does not necessarily follow a logical pattern, but instead requires peer evaluation and other factors (motivational, intuitive, speculative, affective). This historical reconstruction provides another facet of science in the making.

In Chapter 9 participating teachers were asked to provide an overview of the research methodologies of Thomson, Rutherford, Bohr, Millikan and Ehrenhaft, within the context of a philosophy of speculative experiments. One of the participants expressed this in cogent terms:

> in the work of Thomson, Rutherford, Bohr, Millikan and Ehrenhaft besides logic, speculations played an important part … this reconstruction based on the history of science demonstrates that scientists adopt the methodology of idealization (simplifying assumptions) in order to solve complex problems … it is plausible to hypothesize that students adopt similar strategies in order to achieve conceptual understanding.

This response makes interesting reading and following are some of its salient features: (a) recognizes the role of speculations, besides logic; (b) considers the methodology used by these famous scientists as an illustration of idealization (see Chapter 2); and (c) hypothesizes that students may also use a similar strategy, that is "science in the making".

In the traditional science classroom, textbooks and even some science curricula, famous scientists (Newton, Darwin, Maxwell, Einstein) are considered to be geniuses, as they can interpret and understand the experimental data. In Chapter 10 participating teachers were given an opportunity to question this belief. A majority of the teachers agreed that "science does not advance by just having the data" and following is an example of a response by one of the teachers:

> data tend to impose an "imaginary limit" … science progresses, precisely because scientists can foresee beyond the "limit" based on the data … in research on sub-atomic particles the phenomena under study are frequently beyond the limit of detection of the apparatus.

Indeed, this response provides a blue-print of what constitutes "science in the making", namely no amount experimental data can provide conclusive proof for a theory (Duhem–Quine thesis), and thus scientists are forced to use their imagination and creativity in order to explore unknown "territory". This provides yet another facet of science in the making and teaching science as practiced by scientists.

Tentative nature of scientific knowledge is an important marker in the new historical and philosophical landscape, and recognized as such by science educators (Abd-El-Khalick, 2004; Lederman et al., 2002). Chapter 11 provided participating teachers with the opportunity to interpret the different forms of constructivism within this context. All teachers agreed that the present form of constructivism must eventually develop and progress towards other forms of constructivism and following is an example:

> Due to the tentative nature of scientific knowledge—when constructivism does not provide convincing answers for teaching science—when new evidence contrasts with that provided by constructivism—at that stage constructivism will stop being a panacea and undoubtedly must transform into other forms.

This clearly shows how "science in the making" can critique, change, even discard and subsequently facilitate the construction of new theories.

As suggested by Phillips (2005a), all the studies reported in the different chapters of this book were conducted within a framework that represents contemporary history and philosophy of science (Lakatos, Giere, Cartwright, Holton, Laudan). Participating teachers were given an opportunity to discuss various historical episodes with special reference to the postpositivist (Phillips & Burbules, 2000) agenda for educational research. This provided the teachers an innovative experience with respect to the historical reconstruction of scientific progress and facilitated an understanding of "science in the making". Furthermore, teachers were also exposed to the research methodologies of cognitive psychologists (Piaget and Pascual-Leone), who have influenced educational research and practice. This experience was significant as teachers were able to follow and understand the dynamics of scientific progress, based in turn on the evaluation of science textbooks. Having explored "science in the making", teachers were then encouraged to consider an innovative teaching strategy, namely "teaching science as practiced by scientists" and thus go beyond the traditional curricula and textbooks, which are strongly influenced by an inducitivist/positivist philosophy of science. Finally, if the problem precedes the method, and requires a historical reconstruction of the science topic to be introduced in the classroom based on how science is practiced by scientists, it is essential that the methodologists follow the practicing researchers. It is plausible to suggest that such educational practice will receive feedback from the classroom teacher in his role as a researcher and thus facilitate innovative teaching environments.

Notes

6 Qualitative Methodology and Its Pitfalls in Educational Research

1. In order to pursue this further I consulted Yvonna Lincoln, while she was participating as a plenary speaker at the 27 Interamerican Congress of Psychology in Caracas, Venezuela, June 1999. After a lengthy discussion in which she reiterated many of her theoretical assumptions, it became clear that qualitative research can and does use quantitative data. Later she sent me some of her publications with annotations and the following note:

   > Pursuant to our conversation in Caracas, here are pages from our three books, and from our latest chapter. As you can see, we say again and again that: 1) The issue is paradigms, *not* methods and that, 2) Constructivists can and do use both qualitative *and* quantitative methods.
   >
   > (Lincoln, 1999, original italics)

2. Millikan–Ehrenhaft controversy, with respect to the oil drop experiment provides a good example. For details, see Barnes, Bloor and Henry (1996), Holton (1978a) and Niaz (2000a).
3. During class discussions, a brief description of the Millikan–Ehrenhaft controversy with respect to the oil drop experiment was presented. A copy of Niaz (2000a) was facilitated and various participants made photocopies.
4. During class discussions a brief account of Martin Perl's efforts to isolate "quarks" was presented (cf. Perl & Lee, 1997). Perl has now abandoned his active quark research, stating, "Unless I get a splendid idea, I will not continue experimental work in the fractional charge quark field … but I do enjoy thinking about the field and dreaming about its possibilities" (SLAC-PUB-13512, January 2009).

References

Abd-El-Khalick, F. (2004). Over and over again: College students' views of nature of science. In L.B. Flick & N.G. Lederman (Eds.), *Scientific inquiry and nature of science: Implications for teaching, learning, and teacher education* (pp. 389–425). Dordrecht: Kluwer.

Abd-El-Khalick, F. (2005). Developing deeper understanding of nature of science: The impact of a philosophy of science course on preservice science teachers' views and instructional planning. *International Journal of Science Education, 27*, 15–42.

Abd-El-Khalick, F. & Akerson, V.L. (2004). Learning about nature of science as conceptual change: Factors that mediate the development of preservice elementary teachers' views of nature of science. *Science Education, 88*, 785–810.

Abd-El-Khalick, F. & Akerson, V.L. (2007). On the role and use of "theory" in science education research: A response to Johnston, Akerson and Sowell. *Science Education, 91*, 187–194.

Abd-El-Khalick, F. & Lederman, N.G. (2000a). The influence of history of science courses on students' views of nature of science. *Journal of Research in Science Teaching, 37*, 1057–1095.

Abd-El-Khalick, F. & Lederman, N.G. (2000b). Improving science teachers' conceptions of nature of science: a critical review of the literature. *International Journal of Science Education, 22*, 665–701.

Abd-El-Khalick, F., Bell, R.L. & Lederman, N.G. (1998). The nature of science and instructional practice: Making the unnatural natural. *Science Education, 82*, 417–436.

Abd-El-Khalick, F., Boujaoude, S., Duschl, R., Lederman, N., Mamlok, R., Hofstein, A., et al. (2004). Inquiry in science education: International perspectives. *Science Education, 88*, 397–419.

Abd-El-Khalick, F., Waters, M. & Le, A. (2008). Representation of nature of science in high school chemistry textbooks over the past decades. *Journal of Research in Science Teaching, 45*, 835–855.

Achinstein, P. (1987). Scientific discovery and Maxwell's kinetic theory. *Philosophy of Science, 54*, 409–434.

Adamson, S.L., Banks, D., Burtch, M., Cox, F., Judson, E., Turley, J.B., et al. (2003). Reformed undergraduate instruction and its subsequent impact on secondary school teaching practice and student achievement. *Journal of Research in Science Teaching, 40*, 939–957.

Adey, P. & Shayer, M. (1994). *Really raising standards: Cognitive intervention and academic achievement.* London: Routledge.

Adúriz-Bravo, A., Izquierdo, M. & Estany, A. (2002) Una propuesta para estructurar la enseñanza de la filosofía de la ciencia para el profesorado de ciencias en formación. *Enseñanza de las Ciencias, 20*, 465–476.

Akerson, V.L., Morrison, J.A. & McDuffie, A.R. (2006). One course is not enough: Preservice elementary teachers' retention of improved views of nature of science. *Journal of Research in Science Teaching, 43*, 194–213.

Alexander, H.A. (2006). A view from somewhere: Explaining the paradigms of educational research. *Journal of Philosophy of Education, 40*(2), 205–221.

Alters, B.J. (1997). Whose nature of science. *Journal of Research in Science Teaching, 34*, 39–55.

American Association for Physics Teachers, AAPT (1999). What is science? *American Journal of Physics, 67*, 659.

American Association for the Advancement of Science, AAAS (1989). *Project 2061: Science for all Americans.* Washington, DC: AAAS.

American Association for the Advancement of Science, AAAS (1993). *Benchmarks for Science Literacy: Project 2061.* New York: Oxford University Press.

Ander, P. & Sonnessa, A.J. (1981). *Principles of chemistry* (Spanish ed.). New York: Macmillan.

Auerbach, D. (2000). What is science: Isn't there more to it? *American Journal of Physics, 67*, 305.

Barker, P. & Gholson, B. (1984). The history of the psychology of learning as a rational process: Lakatos versus Kuhn. In H.W. Reese (Ed.), *Advances in child development and behavior* (vol. 18, pp. 227–244). New York: Academic Press.

Barnes, B., Bloor, D. & Henry, J. (1996). *Scientific knowledge: A sociological analysis.* Chicago: University of Chicago Press.

Bauer, H.H. (1994). *Scientific literacy and the myth of the scientific method.* Champaign, IL: University of Illinois Press.

Bell, R., Abd-El-Khalick, F., Lederman, N.G., McComas, W.F. & Matthews, M.R. (2001). The nature of science and science education: A bibliography. *Science & Education, 10*, 187–204.

Bereiter, C. (1994). Implications of postmodernism for science, or, science as progressive discourse. *Educational Psychologist, 29*, 3–12.

Beth, E.W. & Piaget, J. (1966). *Mathematical epistemology and psychology.* Dordrecht: Reidel.

Bianchini, J.A. & Colburn, A. (2000). Teaching the nature of science through inquiry to prospective elementary teachers: A tale of two researchers. *Journal of Research in Science Teaching, 37*, 177–209.

Blanco, R. & Niaz, M. (1997). Epistemological beliefs of students and teachers about the nature of science: From "Baconian inductive ascent" to the "irrelevance" of scientific laws. *Instructional Science, 25*, 203–231.

Blanco, R. & Niaz, M. (1998). Baroque tower on a gothic base: A Lakatosian reconstruction of students' and teachers' understanding of structure of the atom. *Science & Education, 7*, 327–360.

Bohr, N. (1913). On the constitution of atoms and molecules. *Philosophical Magazine, 26*, 1–25.

Boring, E.G. (1929). *A history of experimental psychology.* New York: Century.

Borko, H., Liston, D. & Whitcomb, J.A. (2007). Genres of empirical research in teacher education. *Journal of Teacher Education, 58*, 3–11.

Brady, J.E. & Holum, J.R. (1981). *Fundamentals of chemistry* (Spanish ed.). New York: Wiley.

Brady, J.E. & Humiston, J.E. (1996). *General chemistry: Principles and structure* (Spanish ed.). New York: Wiley.

Brainerd, C.J. (1978). The stage question in cognitive developmental theory. *Behavioral and Brain Sciences, 2*, 173–213.

Brito, A., Rodríguez, M.A. & Niaz, M. (2005). A reconstruction of development of the periodic table based on history and philosophy of science and its implications for general chemistry textbooks. *Journal of Research in Science Teaching, 42*, 84–111.

Brown, T. (1994). Creating and knowing mathematics through language and experience. *Educational Studies in Mathematics, 27*, 79–100.

Brown, T.L. & LeMay, H.E. (1988). *Chemistry: The central science* (4th ed., Spanish). Englewood Cliffs, NJ: Prentice Hall.

Brush, S.G. (1974). Should the history of science be rated X? *Science, 18*, 1164–1172.

Brush, S.G. (1976). *The kind of motion we call heat: a history of the kinetic theory of gases in the 19th century*. New York: North-Holland.

Brush, S.G. (1978). Why chemistry needs history and how it can get some. *Journal of College Science Teaching, 7*, 288–291.

Brush, S.G. (2000). Thomas Kuhn as a historian of science. *Science & Education, 9*, 39–58.

Burbules, N.C. & Linn, M.C. (1991). Science education and philosophy of science: congruence or contradiction? *International Journal of Science Education, 13*, 227–241.

Burns, R.A. (1996). *Fundamentals of chemistry* (2nd ed., Spanish). Englewood Cliffs, NJ: Prentice Hall.

Burns, R.B. & Dobson, C.B. (1981). *Statistical tests in experimental psychology research methods and statistics*. Baltimore, MD: University Park Press.

Campanario, J.M. (1999). La ciencia que no enseñamos. *Enseñanza de las Ciencias, 17*, 397–410.

Campanario, J.M. (2002). The parallelism between scientists' and students' resistance to new scientific ideas. *International Journal of Science Education, 24*, 1095–1110.

Campanario, J.M. & Otero, J.C. (2000) Más allá de las ideas previas como dificultades de aprendizaje: las pautas de pensamiento, las concepciones epistemológicas y las estrategias metacognitivas de los alumnos de ciencias. *Enseñanza de las Ciencias, 18*, 155–169.

Campbell, D.T. (1988a). The experimenting society. In E.S. Overman (Ed.), *Methodology and epistemology for social science* (pp. 290–314). Chicago, IL: University of Chicago Press (first published 1971).

Campbell, D.T. (1988b). Can we be scientific in applied social science? In E.S. Overman (Ed.), *Methodology and epistemology for social science* (pp. 315–333). Chicago: University of Chicago Press.

Campbell, D.T. (1988c). Qualitative knowing in action research. In E.S. Overman (Ed.), *Methodology and epistemology for social science* (pp. 360–376). Chicago: University of Chicago Press.

Campbell, D.T. & Stanley, J. (1963). Experimental and quasi-experimental designs for research on teaching. In N.L. Gage (Ed.), *Handbook of research on teaching* (pp. 171–246). Chicago: Rand McNally.

Cardellini, L. (2006). The foundations of radical constructivism: An interview with Ernst Von Glasersfeld. *Foundations of Chemistry, 8*, 177–187.

Carey, S. (1986). Cognitive science and science education. *American Psychologist, 41*, 1123–1130.

Cartwright, N. (1983). *How the laws of physics lie*. Oxford: Clarendon Press.

Cartwright. N. (1989). *Nature's capacities and their measurement*. Oxford: Clarendon Press.

Cartwright, N. (1999). *The dappled world: A study of the boundaries of science*. Cambridge: Cambridge University Press.

Chang, R. (1999). *Chemistry* (6th ed., Spanish). New York: McGraw-Hill.

Chang, S.-N. & Chiu, M.-H. (2008). Lakatos' scientific research programmes as a framework for analyzing informal argumentation about socio-scientific issues. *International Journal of Science Education, 30*, 1753–1773.

Chi, M.T.H. (1992). Conceptual change within and across ontological categories: examples from learning and discovery in science. In R.N. Giere (Ed.) *Cognitive models of science* (pp. 129–186). Minneapolis, MN: University of Minnesota Press.

Chiappetta, E. L., Sethna, G.H. & Fillman, D.A. (1991). A quantitative analysis of high school chemistry textbooks for scientific literacy themes and expository learning aids. *Journal of Research in Science Teaching, 28*, 939–951.

Chinn, C.A. & Brewer, W.F. (1993). The role of anomalous data in knowledge acquisition: A theoretical framework and implications for science instruction. *Review of Educational Research, 63*, 1–49.

Clement, J., Brown, D. & Zietsman, A. (1989). Not all preconceptions are misconceptions: Finding "anchoring conceptions" for grounding instruction on students' intuitions. *International Journal of Science Education, 11*, 554–565.

Clough, M.P. (2006). Learners' responses to the demands of conceptual change: Considerations for effective nature of science instruction. *Science & Education, 15*, 463–494.

Cobern, W.W. (1996). Worldview theory and conceptual change in science education. *Science Education, 80*, 579–610.

Cobern, W.W. & Loving, C.C. (2008). An essay for educators: Epistemological realism really is common sense. *Science & Education, 17*, 425–447.

Cobern, W.W., Gibson, A.T. & Underwood, S.A. (1999). Conceptualizations of nature: An interpretative study of 16 ninth graders' everyday thinking. *Journal of Research in Science Teaching, 36*, 541–564.

Collins, H. (2000). On beyond 2000. *Studies in Science Education, 35*, 169–173.

Collins, H. (2004). *Gravity's shadow: The search for gravitational waves.* Chicago: University of Chicago Press.

Collins, H. (2007). The uses of sociology of science for scientists and educators. *Science & Education, 16*, 217–230.

Collins, H. & Pinch, T. (1993). *The golem: What everyone should know about science.* Cambridge: Cambridge University Press.

Collins, H. & Pinch, T. (1998). *The golem: What you should know about science* (2nd ed.). Cambridge: Cambridge University Press.

Cook, T.D. & Campbell, D.T. (1979). *Quasi-experimentation: Design and analysis for field settings.* Boston: Houghton Mifflin.

Cortéz, R. & Niaz, M. (1999). Adolescents' understanding of *observation, prediction*, and *hypothesis* in everyday and educational contexts. *Journal of Genetic Psychology, 160*, 125–141.

Crowther, J.G. (1910). On the scattering of homogeneous β-rays and the number of electrons in the atom. *Proceedings of the Royal Society*, lxxxiv, 226–247.

Cushing, J.T. (1989). The justification and selection of scientific theories. *Synthese, 78*, 1–24.

Czerniak, C.M. (2009, July). Grand challenges and great opportunities in science education: Is the glass half full or half empty? NARST Presidential Speech. *E-NARST News, 52*(2), pp. 3–8.

Darrigol, O. (2009). A simplified genesis of quantum mechanics. *Studies in History and Philosophy of Modern Physics, 40*, 151–166.

De Berg, K.C. (2003). The development of the theory of electrolytic dissociation: A case study of a scientific controversy and the changing nature of chemistry. *Science & Education, 12*, 397–419.

De Berg, K.C. (2006). The status of constructivism in chemical education research and its relationship to the teaching and learning of the concept of idealization in chemistry. *Foundations of Chemistry, 8*, 153–176.

De Posada, J.M. (1999). Concepciones de los alumnos sobre el enlace químico antes, durante y después de la enseñanza formal: problemas de aprendizaje. *Enseñanza de las Ciencias, 17*, 227–245.

Demastes, S.S., Good, R.G. & Peebles, P. (1995). Students' conceptual ecologies and the process of conceptual change in evolution. *Science Education, 79*, 637–666.

Denzin, N.K. & Lincoln, Y.S. (Eds.) (2000). *Handbook of qualitative research* (2nd ed.). Thousand Oaks, CA: Sage.

Denzin, N.K. & Lincoln, Y.S. (2005). Introduction: The discipline and practice of qualitative research. In N.K. Denzin & Y.S. Lincoln (Eds.), *Sage handbook of qualitative research* (3rd ed., pp. 1–32). Thousand Oaks, CA: Sage.

Dobson, K. (2000). Is physics debatable? *Physics Education, 35*, 1.

Dogan, N. & Abd-El-Khalick, F. (2008). Turkish grade 10 students' and science teachers' conceptions of nature of science: A national study. *Journal of Research in Science Teaching, 45*, 1083–1112.

Dori, Y.J. & Hameiri, M. (2003). Multidimensional analysis system for quantitative chemistry problems: Symbol, macro, micro and process aspects. *Journal of Research in Science Teaching, 40*, 278–302.

Duschl, R.A. (1990). *Restructuring science education: The importance of theories and their development.* New York: Teachers College Press.

Duschl, R.A. & Gitomer, D.H. (1991). Epistemological perspectives on conceptual change: Implications for educational practice. *Journal of Research in Science Teaching, 28*, 839–858.

Dykstra, D.I., Boyle, C.F. & Monarch, I.A. (1992). Studying conceptual change in learning physics. *Science Education, 76*, 615–652.

Ebbing, D.D. (1997). *General chemistry* (5th ed., Spanish). New York: McGraw-Hill.

Eflin, J.T., Glennan, S. & Reisch, G. (1999). The nature of science: A perspective from the philosophy of science. *Journal of Research in Science Teaching, 36*, 107–116.

Ehrenhaft, F. (1910). Uber die kleinsten messbaren elektrizitätsmengen. Zweite vorläufige mitteilung der methode zur bestimmung des elektrischen elementarquantums. *Anzeiger Akad. Wiss* (Vienna), *10*, 118–119.

Ehrenhaft, F. (1941). The microcoulomb experiment. *Philosophy of Science, 8*, 403–457.

Einstein, A. & Infeld, L. (1938/1971). *The evolution of physics.* Cambridge: Cambridge University Press (Original work published in 1938).

Eisner, E.W. (1992). Are all causal claims positivistic? A reply to Schrag. *Educational Researcher, 21*, 8–9.

Eisner, E.W. (1997). The promise and perils of alternative forms of data representation. *Educational Researcher, 26*, 4–10.

Eisner, E.W. (1999). Rejoinder: A response to Tom Knapp. *Educational Researcher, 28*, 19–20.

Eisner, E.W. & Peshkin, A. (Eds.) (1990). *Qualitative inquiry in education: The continuing debate.* New York: Teachers College Press.

Elliot, J. (1985). Facilitating action research in schools: Some dilemmas (chap. 4). In *Action research in education* (Spanish translation, Madrid: Ediciones Morata, 1990).

Ellis, B.D. (1991). Idealization in science. In C. Dilworth (Ed.), *Idealization IV: Intelligibility in science.* Amsterdam: Rodopi.

Erickson, F.E. (1986). Qualitative methods in research on teaching. In M.C. Wittrock (Ed.), *Handbook of research on teaching* (3rd ed., pp. 119–161). New York: Macmillan.

Erickson, F. & Gutierrez, K. (2002). Culture, rigor, and science in educational research. *Educational Researcher, 31*, 21–24.

Eylon, B. & Linn, M.C. (1988). Learning and instruction: An examination of four research perspectives in science education. *Review of Educational Research, 58*, 251–301.

Eysenck, H.J. (1973). *The measurement of intelligence.* Lancaster: Medical & Technical Publishing Co.

Falconer, I. (1987). Corpuscles, electrons, and cathode rays: J.J. Thomson and the "discovery of the electron". *British Journal for the History of Science, 20*, 241–276.

Fernández, I., Gil, D., Carrascosa, J., Cachapuz, A. & Praia, J. (2002). Visiones deformadas de la ciencia transmitidas por la enseñanza. *Enseñanza de las Ciencias, 20*, 477–488.

Feyerabend, P. (1975). *Against method.* London: Verso.

Feynman, R. (1967). *The character of physical law.* Cambridge, MA: MIT Press.

Fleck, L. (1979). *Genesis and development of a scientific fact.* Chicago: University of Chicago Press (first published 1935).

Ford, M. & Wargo, B.M. (2007). Routines, roles, and responsibilities for aligning scientific and classroom practices. *Science Education, 91*, 133–157.

Freidson, E. (1975). *Doctoring together: A study of professional social control.* Chicago: University of Chicago Press.

Freire, P. (1970). *Pedagogy of the oppressed* (chap. 3). New York: Herder & Herder.

Friman, P.C., Allen, K.D., Kerwin, M.L.E. & Larzelere, R. (1993). Changes in modern psychology: A citation analysis of the Kuhnian displacement thesis. *American Psychologist, 48*, 658–664.

Fuller, S. (2000). *Thomas Kuhn: A philosophical history of our times.* Chicago: University of Chicago Press.

Furió, C., Azcona, R. & Guisáosla, J. (2002). Revisión de investigaciones sobre la enseñanza-aprendizaje de los conceptos cantidad de sustancia y mol. *Enseñanza de las Ciencias, 20*, 229–242.

Gage, N.L. (1989). The paradigm wars and their aftermath. *Educational Researcher, 18*, 4–10.

Gage, N.L. (2009). *A conception of teaching.* Dordrecht: Springer.

Galison, P. (1987). *How experiments end.* Chicago: University of Chicago Press.

Gallegos, J.A. (1999). Reflexiones sobre la ciencia y la epistemología científica. *Enseñanza de las Ciencias, 17*, 321–326.

García, J.J. (2000). La solución de situaciones problemáticas: una estrategia didáctica para la enseñanza de la química. *Enseñanza de las Ciencias, 18*, 113–129.

Garfield, E. & Welljams-Dorof, A. (1992). Citation data: Their use as qualitative indicators for science and technology evaluation and policy-making. *Current Contents, 24*, 5–13.

Gaskins, S. (1994). Integrating interpretive and quantitative methods in socialization research. *Merrill-Palmer Quarterly, 40*, 313–333.

Gavroglu, K. (2000). Controversies and the becoming of physical chemistry. In P. Machamer, M. Pera & A. Baltas (Eds.), *Scientific controversies: Philosophical perspectives* (pp. 177–198). New York: Oxford University Press.

Geelan, D. (1997). Epistemological anarchy and the many forms of constructivism. *Science & Education, 6*, 15–28.

Geelan, D. (2006). *Undead theories: Constructivism, eclecticism and research in science education.* Rotterdam: Sense Publishers.

Geiger, H. & Marsden, E. (1909). On a diffuse reflection of the alpha particles. *Proceedings of the Royal Society,* lxxxii. London: Royal Society.

Gholson, B. & Barker, P. (1985). Kuhn, Lakatos and Laudan: Applications in the history of physics and psychology. *American Psychologist, 40*, 755–769.

Giere, R.N. (1988). *Explaining science: A cognitive approach.* Chicago: University of Chicago Press.

Giere, R.N. (1999). *Science without laws*. Chicago: University of Chicago Press.
Giere, R.N. (2006). *Scientific perspectivism*. Chicago: University of Chicago Press.
Gil-Pérez, D., Guisáosla, J., Moreno, A., Cachapuz, A., Pessoa de Carvalho, A.M., Martínez Torregrosa, J., et al. (2002). Defending constructivism in science education. *Science & Education, 11*, 557–571.
Giroux, H.A. (1988). Teachers as transformative intellectuals (chap. 9). In *Teachers as intellectuals: Toward a critical pedagogy of learning*. Boston, MA: Bergin and Garvey.
Glasersfeld, E.V. (1989). Cognition, construction of knowledge, and teaching. *Synthese, 80*, 121–140.
Glasersfeld, E.V. (1992). Constructivism reconstructed: A reply to Suchting. *Science and Education, 1*, 379–384.
Glasson, G.E. & Lalik, R.V. (1993). Reinterpreting the learning cycle from a social constructivist perspective: A qualitative study of teachers' beliefs and practices. *Journal of Research in Science Teaching, 30*, 187–207.
Good, R. (1993). Editorial: The many forms of constructivism. *Journal of Research in Science Teaching, 30*, 1015.
Gooday, G., Lynch, J.M., Wilson, K.G. & Barsky, C.K. (2008). Does science education need the history of science? *Isis, 99*, 322–330.
Gower, B. (1997). *Scientific method: An historical and philosophical introduction*. London: Routledge.
Guba, E.G. (1990). The alternative paradigm dialog. In E.G. Guba (Ed.), *The paradigm dialog* (pp. 17–27). Newbury Park, CA: Sage.
Guba, E.G. & Lincoln, Y.S. (1982). *Effective evaluation*. San Francisco: Jossey-Bass.
Guba, E.G. & Lincoln, Y.S. (1989). *Fourth generation evaluation*. Newbury Park, CA: Sage.
Guba, E.G. & Lincoln, Y.S. (1994). Competing paradigms in qualitative research. In N.K. Denzin & Y.S. Lincoln (Eds.), *Handbook of qualitative research* (pp. 105–117). Thousand Oaks, CA: Sage.
Guba, E.G. & Lincoln, Y.S. (2005). Paradigmatic controversies, contradictions, and emerging confluences. In N.K. Denzin & Y.S. Lincoln (Eds.), *Sage handbook of qualitative research* (3rd ed., pp. 191–215). Thousand Oaks, CA: Sage.
Gunstone, R.F., Gray, C.M. & Searle, P. (1992). Some long-term effects of uninformed conceptual change. *Science Education, 76*, 175–197.
Hanson, N.R. (1958). *Patterns of discovery*. Cambridge: Cambridge University Press.
Heilbron, J.L. & Kuhn, T. (1969). The genesis of the Bohr atom. *Historical Studies in the Physical Sciences, 1*, 211–290.
Hein, M. (1990). *Foundations of college chemistry* (Spanish ed.). Brooks/Cole, Belmont, CA: Brooks/Cole.
Hodson, D. (1985). Philosophy of science, science and science education. *Studies in Science Education, 12*, 25–57.
Hodson, D. (1988a). Towards a Kuhnian approach to curriculum development. *School Organization, 8*, 5–11.
Hodson, D. (1988b). Toward a philosophically more valid science curriculum. *Science Education, 72*, 19–40.
Holton, G. (1969). Einstein, Michelson, and the "crucial" experiment. *Isis, 60*, 133–197.
Holton, G. (1969). Einstein and the "crucial" experiment. *American Journal of Physics, 37*, 968–982.
Holton, G. (1978a). Subelectrons, presuppositions, and the Millikan–Ehrenhaft dispute. *Historical Studies in the Physical Sciences, 9*, 161–224.
Holton, G. (1978b). *The scientific imagination: Case studies*. Cambridge: Cambridge University Press.

Holton, G. (1986). *The advancement of science and its burdens*. Cambridge: Cambridge University Press.

Holton, G. (1992). Ernst Mach and the fortunes of positivism in America. *Isis, 83*, 27–60.

Holton, G. (1993). *Science and anti-science*. Cambridge, MA: Harvard University Press.

Holton, G. (1998). *The scientific imagination*. Cambridge, MA: Harvard University Press.

Hosson, C. & Kaminski, W. (2007). Historical controversy as an educational tool: Evaluating elements of a teaching–learning sequence conducted with the text "Dialogues on the ways that vision operates". *International Journal of Science Education, 29*, 617–642.

Howe, K.R. (1988). Against the quantitative-qualitative incompatibility thesis or dogmas die hard. *Educational Researcher, 17*, 10–16.

Husén, T. (1997). Research paradigms in education. In J.P. Keeves (Ed.), *Educational research, methodology, and measurement: An international handbook* (pp. 16–21). Oxford: Elsevier.

Inhelder, B. & Piaget, J. (1958). *The growth of logical thinking from childhood to adolescence*. New York: Basic Books.

Irwin, A.R. (2000). Historical case studies: Teaching the nature of science in context. *Science Education, 84*, 5–26.

Jenkins, E. (2007) School science: a questionable construct? *Journal of Curriculum Studies, 39*(3), 265–282.

Johnson, J. (1991). Developmental versus language-based factors in metaphor interpretation. *Journal of Educational Psychology, 83*, 470–483.

Johnson, M.A. & Lawson, A.E. (1998). What are the relative effects of reasoning ability and prior knowledge on biology achievement in expository and inquiry classes? *Journal of Research in Science Teaching, 35*, 89–103.

Johnson, R.B. & Christensen, L.B. (2004). *Educational research: Quantitative, qualitative and mixed methods approaches*. Boston, MA: Allyn and Bacon.

Johnson, R.B. & Onwuegbuzie, A.J. (2004). Mixed methods research: A research paradigm whose time has come. *Educational Researcher, 33*, 14–26.

Johnson, R.B. & Turner, L.A. (2003). Data collection strategies in mixed methods research. In A. Tashakori & C. Teddlie (Eds.), *Handbook of mixed methods in social and behavioral research* (pp. 297–319). Thousand Oaks, CA: Sage.

Jones, R.C. (1995) The Millikan oil-drop experiment: Making it worthwhile. *American Journal of Physics, 63*, 970–977.

Justi, R. & Gilbert, J. (1999) A cause of ahistorical science teaching: Use of hybrid models. *Science Education, 83*, 163–177.

Kang, S., Scharmann, L.C. & Noh, T. (2005) Examining students' views on the nature of science: Results from 6th, 8th and 10th graders. *Science Education, 89*, 314–334.

Karmiloff-Smith, A. & Inhelder, B. (1976). If you want to get ahead, get a theory. *Cognition, 3*, 195–212.

Karplus, R., Lawson, A.E., Wollman, W., Appel, M., Bernoff, R., Howe, A., et al. (1977). *Science teaching and the development of reasoning: A workshop*. Berkeley: Regents of the University of California.

Kaufmann, W. (1897). Die magnetische ablenkbarkeit der kathodenstrahlen und ihre abhängigkeit vom entladungspotential. *Annalen de Physik und Chemie, 61*, 544.

Keeves, J.P. & Adams, D. (1997). Comparative methodology in education. In J.P. Keeves (Ed.), *Educational research, methodology, and measurement: An international handbook* (pp. 31–41). Oxford: Elsevier.

Kelly, G.J. (1997). Research traditions in comparative context: A philosophical challenge to radical constructivism. *Science Education, 81*, 355–375.

Kennedy, M.M. (1999). A test of some common contentions about educational research. *American Educational Research Journal, 36*, 511–541.

Kerlinger, F.N. (1975). *Foundations of behavioral research* (2nd ed.). New York: Holt, Rinehart, and Winston.

Kerlinger, F.N. & Lee, H.B. (2002). *Foundations of Behavioral Research* (4th ed., Spanish). New York: McGraw-Hill.

Kesidou, S. & Roseman, J.E. (2002). How well do middle school science programs measure up? Findings from Project 2061's curriculum review. *Journal of Research in Science Teaching, 39*, 522–549.

Khishfe, R. & Lederman, N.G. (2006). Teaching nature of science within a controversial topic: Integrated versus non-integrated. *Journal of Research in Science Teaching, 43*, 395–418.

Kitchener, R.F. (1986). *Piaget's theory of knowledge: Genetic epistemology and scientific reason.* New Haven, CT: Yale University Press.

Kitchener, R.F. (1987). Genetic epistemology, equilibration, and the rationality of scientific change. *Studies in History and Philosophy of Science, 18*, 339–366.

Kitchener, R.F. (1993). Piaget's epistemic subject and science education: Epistemological versus psychological issues. *Science & Education, 2*, 137–148.

Kivinen, O. & Rinne, R. (1998). Methodological challenges for comparative research into higher education. *Interchange, 29*, 121–136.

Klassen, S. (2009). Identifying and addressing student difficulties with the Millikan oil drop experiment. *Science & Education, 18*, 593–607.

Knapp, T.R. (1999). Response to Elliot W. Eisner's "The promise and perils of alternative forms of data representation". *Educational Researcher, 28*, 18–19.

Kousathana, M. & Tsaparlis, G. (2002). Students' errors in solving numerical chemical equilibrium problems. *Chemistry Education: Research and Practice in Europe, 3*, 5–17.

Kuhn, D., Amsel, E. & O'Loughlin, M. (1988). *The development of scientific thinking skills.* San Diego, CA: Academic Press.

Kuhn, T. (1962). *The structure of scientific revolutions.* Chicago: University of Chicago Press.

Kuhn, T.S. (1963). The function of dogma in scientific research. In A.C. Crombie (Ed.), *Scientific change* (pp. 347–395). London: Heinemann.

Kuhn, T.S. (1970). *The structure of scientific revolutions* (2nd ed.). Chicago: University of Chicago Press.

Ladson-Billings, G. & Donnor, J. (2005). The moral activist role of critical race theory scholarship. In N.K. Denzin & Y.S. Lincoln (Eds.), *Sage handbook of qualitative research* (3rd ed., pp. 279–301). Thousand Oaks, CA: Sage.

Lakatos, I. (1970). Falsification and the methodology of scientific research programmes. In I. Lakatos & A. Musgrave (Eds.), *Criticism and the growth of knowledge* (pp. 91–195). Cambridge: Cambridge University Press.

Lakatos, I. (1971). History of science and its rational reconstructions. In R.C. Buck & R.S. Cohen (Eds.), *Boston studies in the philosophy of science* (vol. VIII, pp. 91–136). Dordrecht: Reidel.

Lakatos, I. (1974). The role of crucial experiments in science. *Studies in History and Philosophy of Science, 4*, 309–325.

Lakatos, I. (1999). Lectures on scientific method. In M. Motterlini (Ed.), *For and against method: Including Lakatos's lectures on scientific method and the Lakatos-Feyerabend correspondence* (pp. 19–109). Chicago: University of Chicago Press.

Lakatos, I. & Musgrave, A. (Eds.) (1970). *Criticism and the growth of knowledge.* Cambridge: Cambridge University Press.

Laudan, L. (1977). *Progress and its problems.* Berkeley: University of California Press.

Laudan, R., Laudan, L. & Donovan, A. (1988). Testing theories of scientific change. In A. Donovan, L. Laudan and R. Laudan (Eds.), *Scrutinizing science: Empirical studies of scientific change* (pp. 3–44). Dordrecht: Kluwer.

Lawson, A.E. (1985). A review of research on formal reasoning and science teaching. *Journal of Research in Science Teaching, 22*, 569–617.

Lawson, A.E., McElrath, C.B., Burton, M.S., James, B.D., Doyle, R.P., Woodward, S.L., et al. (1991). Hypothetico-deductive reasoning skill and concept acquisition: testing a constructivist hypothesis. *Journal of Research in Science Teaching, 28*, 953–970.

Lawson, A.E., Reichert, E.A., Costenson, K.L., Fedock, P.M. & Litz, K.K. (1989). Advanced research beyond the ruling theory stage. *Journal of Research in Science Teaching, 26*, 679–686.

Lederman, N.G. (1992). Students' and teachers' conceptions of the nature of science: A review of the research. *Journal of Research in Science Teaching, 29*, 331–359.

Lederman, N.G. (2004). Syntax of nature of science within inquiry and science instruction. In L.B. Flick & N.G. Lederman (Eds.), *Scientific inquiry and nature of science* (pp. 301–317). Dordrecht: Kluwer.

Lederman, N.G., Abd-El-Khalick, F., Bell, R.L. & Schwartz, R.S. (2002). Views of nature of science questionnaire: Toward valid and meaningful assessment of learners' conceptions of nature of science. *Journal of Research in Science Teaching, 39*, 497–521.

Lee, G., Kwon, J., Park, S., Kim, J., Kwon, H. & Park, H. (2003). Development of an instrument for measuring cognitive conflict in secondary-level science classes. *Journal of Research in Science Teaching, 40*, 585–603.

Leite, L. (2002). History of science in science education: Development and validation of a checklist for analyzing the historical content of science textbooks. *Science & Education, 11*, 333–359.

Lewin, K. (1935). The conflict between Aristotelian and Galilean modes of thought in contemporary psychology. In *A dynamic theory of personality* (Selected papers, pp. 1–42). New York: McGraw-Hill.

Lin, H. & Chen, C. (2002). Promoting preservice chemistry teachers' understanding about the nature of science through history. *Journal of Research in Science Teaching, 39*, 773–792.

Lincoln, Y.S. (1989). Trouble in the land: The paradigm revolution in the academic disciplines. In J.C. Smart (Ed.), *Vol. V, Higher education: Handbook of theory and research* (pp. 57–133). New York: Agathon Press.

Lincoln, Y.S. (1990a). Campbell's retrospective and a constructivist's perspective. *Harvard Educational Review, 60*, 501–504.

Lincoln, Y.S. (1990b). Response to "Up from positivism". *Harvard Educational Review, 60*, 508–512.

Lincoln, Y.S. (1999). Personal communication, August 19.

Lincoln, Y.S. & Guba, E.G. (2000). Paradigmatic controversies, contradictions, and emerging confluences. In N.K. Denzin & Y.S. Lincoln (Eds.), *Handbook of qualitative research* (2nd ed.). Thousand Oaks, CA: Sage.

Linn, M.C., Songer, N.B. & Lewis, E.L. (1991). Overview: Students' models and epistemologies of science. *Journal of Research in Science Teaching, 28*, 729–732.

Louden, W. & Wallace, J. (1994). Knowing and teaching science: The constructivist paradox. *International Journal of Science Education, 16*, 649–657.

Loving, C.C. (1997). From the summit of "truth" to the "slippery slopes": Science education's descent through positivist-postmodernist territory. *American Educational Research Journal, 34*, 421–452.

Loving, C.C. & Cobern, W.W. (2000). Invoking Thomas Kuhn: What citation analysis reveals about science education. *Science & Education, 9*, 187–206.

Macbeth, D. (1998). Qualitative methods and the "analytic gaze": An affirmation of scientism? *Interchange, 29*, 137–168.

McComas, W.F. (1996). Ten myths of science: Reexamining what we think we know about the nature of science. *School Science and Mathematics, 96*, 10–16.

McComas, W.F. & Olson, J.K. (1998). The nature of science in international science education standards documents. In W.F. McComas (Ed.), *The nature of science in science education: Rationales and strategies* (pp. 41–52). Dordrecht: Kluwer.

McComas, W.F., Almazroa, H. & Clough, M.P. (1998). The role and character of the nature of science in science education. *Science & Education, 7*, 511–532.

Machamer, P., Pera, M. & Baltas, A. (2000). Scientific controversies: An introduction. In P. Machamer, M. Pera & A. Baltas (Eds.), *Scientific controversies: Philosophical and historical perspectives* (pp. 3–17). New York: Oxford University Press.

McMullin, E. (1985). Galilean idealization. *Studies in History and Philosophy of Science, 16*, 247–273.

Mahan, B.M. & Myers, R.J. (1990). *University chemistry* (4th ed., Spanish). Wilmington, DE: Addison-Wesley.

Malone, M.E. (1993). Kuhn reconstructed: Incommensurability without relativism. *Studies in History and Philosophy of Science, 24*, 69–93.

Marín, N. (1999). Delimitando el campo de aplicación del cambio conceptual. *Enseñanza de las Ciencias, 17*, 80–92.

Marín, N., Benarroch, A. & Gómez, E.J. (2000). What is the relationship between social constructivism and Piagetian constructivism? An analysis of the characteristics of the ideas within both theories. *International Journal of Science Education, 22*, 225–238.

Marín, N., Solano, I. & Jiménez, E. (1999). Tirando del hilo de la madeja constructivista. *Enseñanza de las Ciencias, 17*, 479–492.

Marquit, E. (1978). Philosophy of physics in general physics courses. *American Journal of Physics, 46*, 784–789.

Martin, J. & Sugarman, J. (1993). Beyond methodolatry: Two conceptions of relations between theory and research in research on teaching. *Educational Researcher, 22*, 17–24.

Martínez, A. (1999). Constructivismo radical, marco teórico de investigación y enseñanza de las ciencias. *Enseñanza de las Ciencias, 17*, 493–502.

Martínez, M.M. (1993). Naturaleza y dinámica de los paradigmas científicos (chap. 4). In *El paradigma emergente: Hacia una nueva teoría de la racionalidad* (pp. 52–69). Barcelona: Gedisa.

Martínez, M.M. (1998). *La investigación cualitativa etnográfica en educación* (3rd ed.). México, D.F.: Trillas.

Masterton, W.L., Slowinski, E.J. & Stanitski, C.L. (1985). *Chemical principles* (5th ed., Spanish). Philadelphia: Saunders.

Matthews, M.R. (1987). *Experiment as the objectification of theory: Galileo's revolution. Proceedings of the Second International Seminar on Misconceptions and Educational Strategies in Science and Mathematics* (vol. 1, pp. 289–298). Ithaca, NY: Cornell University.

Matthews, M.R. (1993). Constructivism and science education: Some epistemological problems. *Journal of Science Education and Technology, 2*, 359–370.

Matthews, M.R. (1994a). *Science teaching: The role of history and philosophy of science.* New York: Routledge.

Matthews, M.R. (1994b). Historia, filosofía y enseñanza de las ciencias: la aproximación actual. *Enseñanza de las Ciencias, 12*, 255–277.

Matthews, M.R. (2004). Thomas Kuhn's impact on science education: What lessons can be learned? *Science Education, 88*, 90–118.
Maxwell, J.A. (1990a). Up from positivism. *Harvard Educational Review, 60*, 497–501.
Maxwell, J.A. (1990b). Response to "Campbell's retrospective and a constructivist's perspective". *Harvard Educational Review, 60*, 504–508.
Maxwell, J.A. (1992). Understanding and validity in qualitative research. *Harvard Educational Review, 62*, 279–300.
Maxwell, J.C. (1860). Illustrations of the dynamical theory of gases. *Philosophical Magazine, 19*, 19–32. (*Scientific Papers*, 1965, 377–409, New York: Dover.)
Mayer, M.E. (2000). What is the place of science in educational research? *Educational Researcher, 29*, 38–39.
Mayer, M.E. (2001). Resisting the assault on science: The case for evidence-based reasoning in educational research. *Educational Researcher, 30*, 29–30.
Medawar, P.B. (1967). *The art of the soluble*. London: Methuen.
Mellado, V. (2003). Cambio didáctico del profesorado de ciencias experimentales y filosofía de la ciencia. *Enseñanza de las Ciencias, 21*, 343–358.
Mellado, V., Ruiz, C., Bermejo, M.L. & Jiménez, R. (2006). Contributions from the philosophy of science to the education of science teachers. *Science & Education, 15*, 419–445.
Mendeleev, D. (1869). Ueber die beziehungen der eigenschaften zu den atom gewichtender elemente (C. Giunta, English trans.). *Zeitschrift für Chemie, 12*, 405–406.
Mendeleev, D. (1879). The periodic law of the chemical elements. *The Chemical News, 40*, 1042.
Mendeleev, D. (1889). The periodic law of the chemical elements. *Journal of the Chemical Society, 55*, 634–656 (Faraday lecture, delivered June 4, 1889).
Merton, R.K., Sills, D.L. & Stigler, S.M. (1984). The Kelvin dictum and social science: An excursion into the history of an idea. *Journal of the History of the Behavioral Sciences, 30*, 319–331.
Michell, J. (2003). The quantitative imperative: Positivism, naive realism and the place of qualitative methods in psychology. *Theory & Psychology, 13*, 5–31.
Michell, J. (2005). The meaning of the quantitative imperative: A response to Niaz. *Theory & Psychology, 15*, 257–263.
Millar, R. (1989). Bending the evidence: The relationship between theory and experiments in science education. In R. Millar (ed.), *Doing science: Images of science in science education* (pp. 38–61). London: Falmer Press.
Millar, R. & Driver, R. (1987). Beyond process. *Studies in Science Education, 14*, 33–62.
Millar, R. & Osborne, J.F. (Eds.). (1998). *Beyond 2000: Science education for the future*. London: King's College London.
Miller, S.M., Nelson, M.W. & Moore, M.T. (1998). Caught in the paradigm gap: Qualitative researcher's lived experience and the politics of epistemology. *American Educational Research Journal, 35*, 377–416.
Millikan, R.A. (1910). A new modification of the cloud method of determining the elementary electrical charge and the most probable value of that charge. *Philosophical Magazine, 19*, 209–228.
Millikan, R.A. (1913). On the elementary electrical charge and the Avogadro constant. *Physical Review, 2*, 109–143.
Millikan, R.A. (1916). The existence of a subelectron? *Physical Review, 8*, 595–625.
Mischel, T. (1971). Piaget: Cognitive conflict and the motivation of thought. In T. Mischel (Ed.), *Cognitive development and epistemology* (pp. 311–355). New York: Academic Press.

Mishler, E.G. (1990). Validation in inquiry-guided research: The role of exemplars in narrative studies. *Harvard Educational Review, 60*, 415–442.
Monk, M. and Osborne, J. (1997). Placing the history and philosophy of science on the curriculum: A model for the development of pedagogy. *Science Education*, 81, 405–424.
Montero, M. (1992). Permanencia y cambio de paradigmas en la construcción del conocimiento científico. *Planiuc, 11*, 61–74.
Moreno, L.E. & Waldegg, G. (1998). La epistemología constructivista y la didáctica de las ciencias: ¿Coincidencia o complementaridad? *Enseñanza de las Ciencias, 16*, 421–429.
Moseley, H.G.J. (1913). High frequency spectra of the elements. *Philosophical Magazine, 26*, 1025–1034.
Musgrave, A. (1976). Why did oxygen supplant phlogiston? Research programmes in the chemical revolution. In C. Howson (Ed.), *Method and appraisal in the physical sciences: The critical background to modern science, 1800–1905* (pp. 181–209). Cambridge: Cambridge University Press.
National Research Council, NRC (1996). *National science education standards*. Washington, DC: National Academy Press.
National Research Council, NRC (2002). *Scientific research in education*. Washington, DC: National Academy Press.
Neressian, N.J. (1989). Conceptual change in science and science education. *Synthese, 80*, 163–183.
Newell, A. (1992). Précis of unified theories of cognition. *Behavioral and Brain Sciences, 15*, 425–492.
Niaz, M. (1991a). Role of the epistemic subject in Piaget's genetic epistemology and its importance for science education. *Journal of Research in Science Teaching, 28*, 569–580.
Niaz, M. (1991b). Correlates of formal operational reasoning: A neo-Piagetian analysis. *Journal of Research in Science Teaching, 28*, 19–40.
Niaz, M. (1992). From Piaget's epistemic subject to Pascual-Leone's metasubject: Epistemic transition in the constructivist-rationalist theory of cognitive development. *International Journal of Psychology, 27*, 443–457.
Niaz, M. (1994). Enhancing thinking skills: Domain specific/domain general strategies—A dilemma for science education. *Instructional Science, 22*, 413–422.
Niaz, M. (1995a). Cognitive conflict as a teaching strategy in solving chemistry problems: A dialectic-constructivist perspective. *Journal of Research in Science Teaching, 32*, 959–970.
Niaz, M. (1995b). Progressive transitions from algorithmic to conceptual understanding in student ability to solve chemistry problems: a Lakatosian interpretation. *Science Education, 79*, 19–36.
Niaz, M. (1996a). The controversy between qualitative and quantitative research in education: A legacy of Kuhn's incommensurability thesis. *Perceptual and Motor Skills, 82*, 617–618.
Niaz, M. (1996b). Reasoning strategies of students in solving chemistry problems as a function of developmental level, functional M-capacity and disembedding ability. *International Journal of Science Education, 18*, 525–541.
Niaz, M. (1996c). How students circumvent problem-solving strategies that require greater cognitive complexity. *Journal of College Science Teaching, 25*, 361–363.
Niaz, M. (1997). Can we integrate qualitative and quantitative research in science education? *Science & Education, 6*, 291–300.
Niaz, M. (1998). From cathode rays to alpha particles to quantum of action: a rational reconstruction of structure of the atom and its implications for chemistry textbooks. *Science Education, 82*, 527–552.

Niaz, M. (1999a). The role of idealization in science and its implications for science education. *Journal of Science Education and Technology, 8*, 145–150.
Niaz, M. (1999b). Should we put observations first? *Journal of Chemical Education, 76*, 734.
Niaz, M. (2000a). The oil drop experiment: A rational reconstruction of the Millikan–Ehrenhaft controversy and its implications for chemistry textbooks. *Journal of Research in Science Teaching, 37*, 480–508.
Niaz, M. (2000b). A rational reconstruction of the kinetic molecular theory of gases based on history and philosophy of science and its implications for chemistry textbooks. *Instructional Science, 28*, 23–50.
Niaz, M. (2000c) A framework to understand students' differentiation between heat energy and temperature and its educational implications. *Interchange, 31*, 1–20.
Niaz, M. (2001a). Understanding nature of science as progressive transitions in heuristic principles. *Science Education, 85*, 684–690.
Niaz, M. (2001b). A rational reconstruction of the origin of the covalent bond and its implications for general chemistry textbooks. *International Journal of Science Education, 23*, 623–641.
Niaz, M. (2001c). How important are the laws of definite and multiple proportions in chemistry and teaching chemistry? A history and philosophy of science perspective. *Science & Education, 10*, 243–266.
Niaz, M. (2001d). Constructivismo social: Panacea o problema? *Interciencia, 26*, 185–189.
Niaz, M. (2002). Facilitating conceptual change in students' understanding of electrochemistry. *International Journal of Science Education, 24*, 425–439.
Niaz, M. (2003). The oil drop experiment: How did Millikan decide what was an appropriate drop? *Alberta Journal of Educational Research, 49*, 368–374.
Niaz, M. (2004a). Exploring alternative approaches to methodology in educational research. *Interchange*, 35, 155–184.
Niaz, M. (2004b). Did Columbus *hypothesize* or *predict* that if he sailed due West, he would arrive at the Indies? *Journal of Genetic Psychology, 165*, 149–156.
Niaz, M. (2005a). The quantitative imperative vs the imperative of presuppositions. *Theory & Psychology, 15*, 247–256.
Niaz, M. (2005b). An appraisal of the controversial nature of the oil drop experiment: Is closure possible? *British Journal for the Philosophy of Science, 56*, 681–702.
Niaz, M. (2006). Facilitating chemistry teachers' understanding of alternative interpretations of conceptual change. *Interchange, 37*(1–2), 129–150.
Niaz, M. (2007). Can findings of qualitative research in education be generalized? *Quality and Quantity: International Journal of Methodology, 41*, 429–445.
Niaz, M. (2008a). Do we need to write physical science textbooks within a history and philosophy of science perspective? In M.V. Thomase (Ed.), *Science education in focus* (pp. 15–65). New York: Nova Science Publishers.
Niaz, M. (2008b). What "ideas-about-science" should be taught in school science? A chemistry teachers' perspective. *Instructional Science, 36*, 233–249.
Niaz, M. (2008c). *Teaching general chemistry: A history and philosophy of science approach.* New York: Nova Science Publishers.
Niaz, M. (2008d). A rationale for mixed methods (integrative) research programmes in education. *Journal of Philosophy of Education, 42*(2), 287–305.
Niaz, M. (2008e). Whither constructivism? A chemistry teachers' perspective. *Teaching and Teacher Education, 24*, 400–416.
Niaz, M. (2009a). *Critical appraisal of physical science as a human enterprise: Dynamics of scientific progress.* Dordrecht: Springer.

Niaz, M. (2009b). Progressive transitions in chemistry teachers' understanding of nature of science based on historical controversies. *Science & Education, 18*, 43–65.

Niaz, M. (2009c). Qualitative methodology and its pitfalls in educational research. *Quality and Quantitiy: International Journal of Methodology, 43*, 535–551.

Niaz, M. (2010). Are we teaching science as practiced by scientists? *American Journal of Physics, 78*(1), 5–6.

Niaz, M. & Cardellini, L. (2010). What can the Bohr–Sommerfeld model show students of chemistry in the 21st century? *Journal of Chemical Education, 87.*

Niaz, M. (2010). Science curriculum and teacher education: The role of presuppositions, contradictions, controversies and speculations vs Kuhn's "normal science". *Teaching and Teacher Education*, 26, 891–899.

Niaz, M. & Chacón, E. (2003). A conceptual change teaching strategy to facilitate high school students' understanding of electrochemistry. *Journal of Science Education and Technology, 12*, 129–134.

Niaz, M. & Rodríguez, M.A. (2002). Improving learning by discussing controversies in 20th century physics. *Physics Education, 37*, 59–63.

Niaz, M., Abd-El-Khalick, F., Benarroch, A., Cardellini, L., Laburú, C.E., Marín, N., et al. (2003). Constructivism: Defense or a continual critical appraisal—A response to Gil-Pérez et al. *Science & Education, 12*, 787–797.

Niaz, M., Aguilera, D., Maza, A. & Liendo, G. (2002). Arguments, contradictions, resistances, and conceptual change in students' understanding of atomic structure. *Science Education, 86*, 505–525.

Niaz, M., Klassen, S., McMillan, B. & Metz, D. (2010a). Leon Cooper's perspective on teaching science: An interview study. *Science & Education, 19*, 39–54.

Niaz, M., Klassen, S., McMillan, B. & Metz, D. (2010b). Reconstruction of the history of the photoelectric effect and its implications for general physics textbooks. *Science Education, 94*, in press.

Niaz, M., Rodríguez, M.A. & Brito, A. (2004). An appraisal of Mendeleev's contribution to the development of the periodic table. *Studies in History and Philosophy of Science, 35*, 271–282.

Nola, R. (1997). Constructivism in science and science education: A philosophical critique. *Science & Education, 6*, 55–83.

Novak, J.D. (1977). An alternative to Piagetian psychology for science and mathematics education. *Science Education, 61*, 453–477.

Onwuegbuzie, A.J. & Leech, N.L (2005). Taking the "Q" out of research: Teaching research methodology courses without the divide between qualitative and quantitative paradigms. *Quality & Quantity: International Journal of Methodology, 39*, 267–296.

Osborne, J.F. (1996). Beyond constructivism. *Science Education, 80*, 53–80.

Osborne, J.F. (2007). Science education for the twenty first century. *Eurasia Journal of Mathematics, Science and Technology Education, 3*(3), 173–184.

Osborne, J.F. & Collins, S. (2001). Pupils' views of the role and value of the science curriculum: A focus-group study. *International Journal of Science Education, 23*(5), 441–468.

Osborne, J., Collins, S., Ratcliffe, M., Millar, R. & Duschl, R. (2003). What "ideas-about-science" should be taught in school science? A Delphi study of the expert community. *Journal of Research in Science Teaching, 40*, 692–720.

Oulton, C., Dillon, J. & Grace, M.M. (2004). Reconceptualizing the teaching of controversial issues. *International Journal of Science Education, 26*, 411–423.

Padilla, K. & Furio-Mas, C. (2008). The importance of history and philosophy of science in correcting distorted views of "amount of substance" and "mole" concepts in chemistry teaching. *Science & Education, 17*, 403–424.

Paul, J.L. & Marfo, K. (2001). Preparation of educational researchers in philosophical foundations of inquiry. *Review of Educational Research, 71*, 525–547.

Pascual-Leone, J. (1970). A mathematical model for the transition rule in Piaget's developmental stages. *Acta Psychologica, 32*, 301–345.

Pascual-Leone, J. (1978). Compounds, confounds, and models in developmental information processing: A reply to Trabasso and Foellinger. *Journal of Experimental Child Psychology, 26*, 18–40.

Pascual-Leone, J. (1987). Organismic processes for neo-Piagetian theories: A dialectical causal account of cognitive development. *International Journal of Psychology, 22*, 531–570.

Pascual-Leone, J. (1988). Affirmations and negations, disturbances and contradictions, in understanding Piaget: Is his later theory causal? *Contemporary Psychology, 33*, 420–421.

Perkins, D.N. (1999). The many faces of constructivism. *Educational Leadership, 56*, 6–11.

Perkins, D.N. (2006). Constructivism and troublesome knowledge. In J.H.F. Meyer & R. Land (Eds.), *Overcoming barriers to student understanding: Threshold concepts and troublesome knowledge* (pp. 33–47). London: Routledge.

Perl, M.L. (2005). Personal communication (email) to the author, December 1.

Perl, M.L. & Lee, E.R. (1997). The search for elementary particles with fractional electric charge and the philosophy of speculative experiments. *American Journal of Physics, 65*, 698–706.

Perrin, C.E. (1988). The chemical revolution: Shifts in guiding assumptions. In A. Donovan, L. Laudan & R. Laudan (Eds.), *Scrutinizing science: Empirical studies of scientific change* (pp. 105–124). Dordrecht: Kluwer.

Peshkin, A. (2000). The nature of interpretation in qualitative research. *Educational Researcher, 29*, 5–9.

Petrucci, D. & Dibar, M.C. (2001). Imagen de la ciencia en alumnos universitarios: Una revisión y resultados. *Enseñanza de las Ciencias, 19*, 217–229.

Phillips, D.C. (1983). After the wake: Postpositivistic educational thought. *Educational Researcher, 12*, 4–12.

Phillips, D.C. (1987). Validity in quality research: Why the worry about warrant will not wane. *Education and Urban Society, 20*, 9–24.

Phillips, D.C. (1990). Subjectivity and objectivity: An objective inquiry. In E.W. Eisner & A. Peshkin, A. (Eds.), *Qualitative inquiry in education: The continuing debate*. New York: Teachers College Press.

Phillips, D.C. (1994a). Positivism, antipositivism and empiricism. In T. Husén & T.N. Postlethwaite (eds.), *The international encyclopedia of education* (2nd ed., pp. 4630–4634). Oxford: Pergamon.

Phillips, D.C. (1994b). Telling it straight: Issues in assessing narrative research. *Educational Psychologist, 29*, 13–21.

Phillips, D.C. (1995). The good, the bad, and the ugly: The many faces of constructivism. *Educational Researcher, 24*, 5–12.

Phillips, D.C. (2005a). The contested nature of empirical educational research (and why philosophy of education offers little help). *Journal of Philosophy of Education, 39*(4), 577–597.

Phillips, D.C. (2005b). The contested nature of scientific educational research: A guide for the perplexed. Keynote address at the 11th Biennial European Conference for Research on Learning and Instruction, Nicosia, Cyprus (available online: http://earli2005conference.ac.cy).

Phillips, D.C. (2006). A guide for the perplexed: Scientific educational research, methodolatry, and the gold versus platinum standards. *Educational Research Review, 1*(1), 15–26.

Phillips, D.C. & Burbules, N.C. (2000). *Postpositivism and educational research.* New York: Rowman & Littlefield.

Piaget, J. (1985). *The equilibration of cognitive structures: The central problem of intellectual development.* Chicago: University of Chicago Press.

Piaget, J. & Garcia, R. (1989). *Psychogenesis and the history of science.* New York: Columbia University Press.

Pickering, M. (1990). Further studies on concept learning versus problem solving. *Journal of Chemical Education, 67*, 254–255.

Pintrich, P.R., Marx, R.W. & Boyle, R.A. (1993). Beyond cold conceptual change: The role of motivational beliefs and classroom contextual factors in the process of conceptual change. *Review of Educational Research, 63*, 167–199.

Pocoví, M.C. (2007). The effects of a history based instructional material on the students' understanding of field lines. *Journal of Research in Science Teaching, 44*, 107–132.

Polanyi, M. (1964). *Personal knowledge: Towards a post-critical philosophy.* New York: Harper & Row (first published 1958).

Pomeroy, D. (1993). Implications of teachers' beliefs about the nature of science: Comparison of the beliefs of scientists, secondary science teachers, and elementary teachers. *Science Education, 77*, 261–278.

Pomeroy, D. (2003). Implications of teachers' beliefs about the nature of science. *Science Education, 77*, 261–278.

Popper, K.R. (1970). Normal science and its dangers. In I. Lakatos & A. Musgrave (Eds.), *Criticism and the growth of knowledge* (pp. 51–59). Cambridge: Cambridge University Press.

Posner, G.J., Strike, K.A., Hewson, P.W. & Gertzog, W.A. (1982). Accommodation of a scientific conception: Toward a theory of conceptual change. *Science Education, 66*, 211–227.

Quine, W.V.O. (1953). *From a logical point of view.* New York: Harper & Row.

Ratnesar, N. & Mackenzie, J. (2006). The quantitative-qualitative distinction and the null hypothesis significance testing procedure. *Journal of Philosophy of Education, 40*(4), 501–509.

Reese, H.W. & Overton, W.F. (1972). On paradigm shifts. *American Psychologist, 27*, 1197–1199.

Rigden, J.S. & Stuewer, R.H. (2005). Do physicists understand physics? *Physics in Perspective, 7*, 387–389.

Rodríguez, M.A. & Niaz, M. (2004a). A reconstruction of structure of the atom and its implications for general physics textbooks. *Journal of Science Education and Technology, 13*, 409–424.

Rodríguez, M.A. & Niaz, M. (2004b). The oil drop experiment: An illustration of scientific research methodology and its implications for physics textbooks. *Instructional Science, 32*, 357–386.

Rodríguez, M.A. & Niaz, M. (2004c). La teoría cinético-molecular de los gases en libros de física: Una perspectiva basada en la historia y filosofía de la ciencia. *Revista de Educación en Ciencias, 5*, 68–72.

Russo, S. & Silver, M. (2002). *Introductory chemistry* (2nd ed.). San Francisco: Benjamin Cummings.

Rutherford, E. (1911). The scattering of alpha and beta particles by matter and the structure of atom. *Philosophical Magazine, 21*, 669–688.

Rutherford, E. (1915). The constitution of matter and the evolution of the elements. Address to the Annual Meeting of the National Academy of Sciences (pp. 167–202). Washington, DC: Smithsonian Institution.

Rutherford, E. & Geiger, H. (1908). The charge and the nature of the alpha particle. *Proceedings of the Royal Society, 81*, 168–171.

Sadler, T.D., Chambers, F.W. & Zeidler, D.L. (2004). Student conceptualizations of the nature of science in response to a socio-scientific issue. *International Journal of Science Education, 26*, 387–409.

Sale, J.E.M. & Brazil, K. (2004). A strategy to identify critical appraisal criteria for primary mixed-methods studies. *Quality & Quantity: International Journal of Methodology, 38*, 351–365.

Saloman, G. (1991). Transcending the qualitative-quantitative debate: The analytic and systemic approaches to educational research. *Educational Researcher, 20*, 10–18.

Sanger, M.J. & Greenbowe, T.J. (1997) Common student misconceptions in electrochemistry: galvanic, electrolytic and concentration cells. *Journal of Research in Science Teaching, 34*, 377–398.

Scharmann, L.C. & Smith, M.U. (2001). Further thoughts on defining versus describing the nature of science: A response to Niaz. *Science Education, 85*, 691–693.

Schrag, F. (1992). In defense of positivist research paradigms. *Educational Researcher, 21*, 5–8.

Schwab, J.J. (1962). *The teaching of science as enquiry*. Cambridge, MA: Harvard University Press.

Schwab, J.J. (1974). The concept of the structure of a discipline. In E.W. Eisner and E. Vallance (Eds.), *Conflicting conceptions of curriculum* (pp. 162–175). Berkeley: McCutchan.

Segre, M. (1989). Galileo, Viviani, and the tower of Pisa. *Studies in History and Philosophy of Science, 20*, 435–451.

Sensevy, G., Tiberghein, A., Santini, J., Laubé, S. & Griggs, P. (2008). An epistemological approach to modeling: Case studies and implications for science teaching. *Science Education, 92*, 424–446.

Shapin, S. (1996). *The scientific revolution*. Chicago: University of Chicago Press.

Shayer, M. & Adey, P. (1981). *Towards a science of science teaching*. London: Heinemann.

Shim, S.H. (2008). A philosophical investigation of the role of teachers: A synthesis of Plato, Confucius, Buber, and Freire. *Teaching and Teacher Education, 24*, 515–535.

Shulman, L.S. (1986). Paradigms and research programs in the study of teaching: A contemporary perspective. In M.C. Wittrock (Ed.), *Handbook of research on teaching* (3rd ed., pp. 3–36). New York: Macmillan.

Siegel, H. (1978). Kuhn and Schwab on science texts and the goals of science education. *Educational Theory, 28*, 302–309.

Siegel, H. (1979). On the distortion of the history of science in science education. *Science Education, 63*, 111–118.

Sienko, M.J. & Plane, R.A. (1971). *Chemistry* (4th ed., Spanish). New York: McGraw-Hill.

Slater, M. (2008). How to justify teaching false science. *Science Education, 92*, 526–542.

Smeyers, P. (2006). "What it makes sense to say": Education, philosophy and Peter Winch on social science. *Journal of Philosophy of Education, 40*(4), 463–485.

Smith, M.U. & Scharmann, L.C. (1999). Defining versus describing the nature of science: A pragmatic analysis for classroom teachers and science educators. *Science Education, 83*, 493–509.

Smith, M.U., Lederman, N.G., Bell, R.L., McComas, W.F. & Clough, M.P. (1997). How great is the disagreement about the nature of science: A response to Alters. *Journal of Research in Science Teaching, 34*, 1101–1103.

Solbes, J. & Traver, M.J. (1996). La utilización de la historia de las ciencias en la enseñanza de la física y química. *Enseñanza de las Ciencias, 14*, 103–112.

Solbes, J. & Traver, M.J. (2001). Resultados obtenidos introduciendo historia de la ciencia en las clases de física y química: mejora de la imagen de la ciencia y desarrollo de actitudes positivas. *Enseñanza de las Ciencias, 19*, 151–162.

Solomon, J. (1994). The rise and fall of constructivism. *Studies in Science Education, 23*, 1–19.

Solomon, J., Scott, L. & Duveen, J. (1996). Large scale exploration of pupils' understanding of the nature of science. *Science Education, 80*, 493–508.

Southerland, S.A., Johnston, A. & Sowell, S. (2006). Describing teachers' conceptual ecologies for the nature of science. *Science Education, 90*, 874–906.

Sowell, S., Johnston, A. & Southerland, S. (2007). Calling for a focus on where learning happens: A response to Abd-El-Khalick and Akerson. *Science Education, 91*, 195–199.

Spearman, C. (1937). *Psychology down the ages* (vol. 1). London: Macmillan.

Stinner, A. (1992). Science textbooks and science teaching: From logic to evidence. *Science Education, 76*, 1–16.

Stinner, A. & Teichman, J. (2003). Lord Kelvin and the age-of-the-earth debate: A dramatization. *Science & Education, 12*, 213–228.

Strike, K.A. & Posner, G.J. (1992). A revisionist theory of conceptual change. In R.A. Duschl & R.J. Hamilton (Eds.), *Philosophy of science, cognitive psychology, and educational theory and practice* (pp. 147–176). Albany, NY: State University of New York Press.

Suchting, W.A. (1992). Constructivism deconstructed. *Science & Education, 1*, 223–254.

Taber, K.S. (2001). Shifting sands: A case study of conceptual development as competition between alternative conceptions. *International Journal of Science Education, 23*, 731–753.

Taber, K.S. (2006). Constructivism's new clothes: The trivial, the contingent, and a progressive research programme into the learning of science. *Foundations of Chemistry, 8*, 189–219.

Tashakkori, A. & Teddlie, C. (2003). Preface. In A. Tashakkori & C. Teddlie (Eds.), *Handbook of mixed methods in social and behavioral research* (pp. ix–xv). Thousand Oaks, CA: Sage.

Taylor, S.J. & Bogdan, R. (1984). Participant observation in the field (chap. 3). In *Introduction to qualitative research methods: The search for meanings*. New York: Wiley.

Teddlie, C. & Tashakkori, A. (2003). Major issues and controversies in the use of mixed methods in the social and behavioral sciences. In A. Tashakkori & C. Teddlie (Eds.), *Handbook of mixed methods in social and behavioral research* (pp. 3–49). Thousand Oaks, CA: Sage.

Thagard, P. (1990). The conceptual structure of the chemical revolution. *Philosophy of Science, 57*, 183–209.

Thagard, P. (1992). *Conceptual revolutions*. Princeton, NJ: Princeton University Press.

Thomson, J.J. (1897). Cathode rays. *Philosophical Magazine, 44*, 293–316.

Thomson, W. (1891). *Popular lectures and addresses* (vol. 1). London: Macmillan.

Tobias, S. (1993). What makes science hard? A Karplus lecture. *Journal of Science Education and Technology, 2*, 297–304.

Tobias, S. & Duffy, T.M. (2009). Editors. *Constructivist instruction: Success or failure?* New York: Routledge.

Tobin, K. & LaMaster, S.U. (1995). Relationships between metaphors, beliefs, and actions in a context of science curriculum change. *Journal of Research in Science Teaching, 32*, 225–242.

Toulmin, S. (1958). *The uses of argument*. Cambridge: Cambridge University Press.

Toulmin, S. (1961). *Foresight and understanding*. Bloomington, IN: Indiana University Press.

Tsai, C.-C. (2002). Nested epistemologies: Science teachers' beliefs of teaching, learning and science. *International Journal of Science Education, 24*, 771–783.

Tsai, C.-C. (2003). Taiwanese science students' and teachers' perceptions of the laboratory learning environments: Exploring epistemological gaps. *International Journal of Science Education, 25*, 847–860.

Tsai, C.-C. (2006). Reinterpreting and reconstructing science: Teachers' view changes toward the nature of science by courses of science education. *Teaching and Teacher Education, 22*, 363–375.

Tsai, C.-C. (2007). Teachers' scientific epistemological views: The coherence with instruction and students' views. *Science Education, 91*, 222–243.

Van Aalsvoort, J. (2004). Logical positivism as a tool to analyse the problem of chemistry's lack of relevance in secondary school chemical education. *International Journal of Science Education, 26*, 1151–1168.

Van Berkel, B., De Vos, W., Verdonk, A.H. & Pilot, A. (2000). Normal science education and its dangers: The case of school chemistry. *Science & Education, 9*(1–2), 123–159.

Vaquero, J., Rojas de Astudillo, L. & Niaz, M. (1996). Pascual-Leone and Baddeley's models of information processing as predictors of academic performance. *Perceptual and Motor Skills, 82*, 787–798.

Von Aufschnaiter, C., Erduran, S., Osborne, J. & Simon, S. (2008). Arguing to learn and learning to argue: Case studies of how students' argumentation relates to their scientific knowledge. *Journal of Research in Science Teaching, 45*, 101–131.

Vosniadou, S. (1994). Capturing and modeling the process of conceptual change. *Learning and Instruction, 4*, 45–69.

Vuyk, R. (1981). *Overview and critique of Piaget's genetic epistemology 1965–1980.* New York: Academic Press.

Weinberg, S. (2001). Physics and history. In J.A. Labinger & H.M. Collins (Eds.), *The one culture: A conversation about science* (pp. 116–127). Chicago: University of Chicago Press.

White, B.Y. (1993). Thinker tools: Causal models, conceptual change, and science education. *Cognition and Instruction, 10*, 1–100.

Whitten, K.W., Davis, R.E. & Peck, M.L. (1998). *General chemistry* (3rd ed., Spanish). New York: McGraw-Hill.

Wiechert, E. (1897). Ergebniss einer messung der geschwindigkeit der kathodenstrahlen. *Schriften der Physicalischökonomisch Gesellschaft zu Königsberg, 38*, 3.

Wilson, D. (1983). *Rutherford: Simple genius.* Cambridge, MA: MIT Press.

Wilson, K.G. & Barsky, C.K. (1998). Applied research and development: Support for continuing improvement in education. *Daedalus, 127*, 233–258.

Windschitl, M. (2004). Folk theories of "inquiry": How preservice teachers reproduce the discourse and practices of atheoretical scientific method. *Journal of Research in Science Teaching, 41*(5), 481–512.

Windschitl, M., Thompson, J. & Braaten, M. (2008). Beyond the scientific method: Model-based inquiry as a new paradigm of preference for school science investigations. *Science Education, 92*, 941–967.

Wong, S.L., Hodson, D., Kwan, J., Wai Jung, B.H. (2008). Turning crisis into opportunity: Enhancing student-teachers' understanding of nature of science and scientific inquiry through a case study of the scientific research in severe acute respiratory syndrome. *International Journal of Science Education, 30*(11), 1417–1439.

Zoller, U. & Tsaparlis, G. (1997). Higher and lower-order cognitive skills: The case of chemistry. *Research in Science Education, 27*, 117–130.

Index

Note: Pages numbers in *italics* denote tables.

Abd-El-Khalick, F. 5, 6, 22, 28, 152, 170, 175, 186
ACEPT (Arizona Collaborative for Excellence in the Preparation of Teachers) program 2
Achinstein, P. 27
Adams, D. 81
Adey, P. 108
administrative policies 3
Adúriz-Bravo, A. 115, 130
Alexander, H.A. 36
algorithmic learning 12
algorithms 10, 102, 103, 124, 125, 135, 136, 146, 190
Allen, K.D. 45
alpha particles, single/compound scattering 7, 8, 23–4, 55, 116, 124, 134, 144, 170, 197
alternative interpretations: of conceptual change 11, 113–25, 190, 192; of data ix, 1, 3, 5, 6, 9, 32, 97, 124, 127, 138, 144, 155–6, 164, 174, 187, 189, 190, 192, 197
American Association for the Advancement of Science ix, 19
American Association of Physics Teachers 149
American Journal of Physics 146
American Physical Society 149
Amsel, E. 108
analogies, cross-domain 36
analysis and interpretation of data 150, 160, 164; *see also* experimental data (empirical evidence), interpretation of
anomalies 14, 23, 35–6, 49
argumentation 16
arguments 3, 33, 44, 86, 87, 127
Aristotelian idealization 102, 189
Aristotelian physics 25, 36
Aristotle 13
Arizona Collaborative for Excellence in the Preparation of Teachers (ACEPT) program 2
assumptions: guiding 1, 9, 23, 32, 33, 35, 128, 188; simplifying 14, 15, 125, 139, 146, 197
astronomy, Copernican 25, 36
atomic models/theories 8, 36, 54–5, 165, *167*, 186; Bohr 5, 24–5, 28, 30–1, 36, 55, 116, 134, 139, 144, 147, *167*, 192–3; Bohr-Sommerfeld 193; Chadwick *167*; Dalton 28, 40, 55; De Broglie *167*; Gell-Mann *167*; Heisenberg *167*; heuristic power of 26–7; hybrid 28; Pauli *167*; paradoxical stability of Rutherford atom 30–1; periodicity and 40–1; Perl *167*; Rutherford 5, 23, 26, 28, 30–1, 55, 116, 134, 139, 144, 161–2, *167*; Schrödinger *167*; Sommerfeld *167*, 193; Thomson 5, 24, 26, 28, 30, 55, 134, 139, 161–2, *167*
Ausubelian constructivism 166, *167*
authenticity 10, 75, 85, 88, 90, 98–9, 101, 189
auxiliary hypotheses 15

Bacon, F. 151
Baltas, A. 21, 182
Barker, P. 36, 37, 46
Barnes, B. 26, 27
Barsky, C.K. 5, 16
behavioral objectives 56–7, 58, 188
behavioral psychology 45
behaviorism 12
Benchmarks for Science Literacy 19
Berthelot, M. 40
Beyond 2000 31, 20
biology 150
Blanco, R. 128, 152, 170, 174
Bloor, D. 26
boat metaphor (Putnam) 176
Bogdan, R. 50, 66
Bohr, N. 28, 40, 114, 132, 140; four postulates 15, 137, 144, 147, 190, 192–3; model of the atom 5, 28, 30–1, 36, 55, 116, 134, 139, 144, 147, *167*, 192–3 (inconsistent foundations

25; as representing a deep philosophical chasm 25); quantum of action postulate 25, 116, 124, 134, 170; research program 15, 16
Boring, E.G. 43
Borko, H. 1–2
Boyle's law 27
Braaten, M. 165
Brainerd, C.J. 73
Brazil, K. 46
Brewer, W.F. 122, 123
Brito, A. 28
Broglie, M. de *167*
Brown, T. 50, 74, 89, 169
Brush, S.G. 16, 27, 117, 147
Burbules, N.C. 34, 140, 187, 198
Burns, R.B. 103, 107, 112

caloric theory of heat 114, 117, 197
Campanario, J.M. 115, 117–18, 119–20, 130, 152, 170
Campbell, D.T. 6, 45, 50, 54, 65, 66, 95, 98
cancer and smoking 109
Cannizaro, S. 40
Cardellini, L. 186, 193
Carey, S. 73
Cartwright, N. 4, 9, 13, 35, 37, 38, 48
cathode ray experiments 7, 8, 23, 30, 41–2, 86–7, 128–9, 134
causal relations, postulation of 99, 100, 101, 102
Center of Genetic Epistemology 72
certainty, science and 150, *160*
ceteris paribus clauses 14, 37, 101, 102
Chadwick, J. *167*
Chang, S.-N. 16
chemical elements, periodic law of 27–8, 40–1
Chemical Revolution 39
chemistry 150; empirical nature of 127, 136, *142–3*, 145–6
chemistry curriculum 11, 12–13
chemistry education 13–14; myths in 131–2, 136–7, 147; and philosophy of speculative experiments 139–41
chemistry teachers: elaboration of hypothesis and prediction *106*, *111*; understanding of conceptual change 114–25; understanding of constructivism 169–86; understanding of nature of science 11, 129–48, 151, 190, 192–3
chemistry textbooks 8, 25, 27, 28–9, 144; and heuristic principles 129; representation of nature of science (NOS) 28; representation of scientific method in 124, 132, 137–9, 146, 192, 193; role of controversies in 23, 24
Chen, C. 152, 170, 174
Chinn, C.A. 122, 123
Chiu, M.-H. 16
citation analysis, of Kuhnian displacement thesis 45
classic positivism 12
classical experimentalism 47
classical mechanics 4
Clough, M.P. 126
Cobern, W.W. 4, 5–6, 168, 169
cognitive abilities 6
cognitive conflicts 57–8, 121, 122, 123, 125
cognitive psychology 9, 37, 43–4, 45, 48, 72–3, 186, 193, 196, 198
collaboration and cooperation 150, *160*, 161
Collins, H. 9, 31, 33, 196; trilemma 20–1, 30, 32, 157–9, 159, 164, 191
Columbus' discovery of America 10, 104–9
competitive cross-validation 6, 8
complexity of theory construction 151, 155–6
comprehension, teaching for 57
Comte, A. 12
concept maps 166
conceptual change: alternative interpretations of 11, 113–25, 190, 192; cognitive conflicts and 121, 122, 123, 125; constructivism and 63; resistance to 119–20, 123
conceptual understanding 21
conflicting frameworks 31, 32
constructive model of falling bodies 13
constructivism 9, 10, 11–12, 53, 87, 146, 165, 166–86, 191, 194–5; Ausubelian 166; and boat metaphor of Putnam 176; and conceptual change 63; critical appraisal of 63–4, 69–70, 166, 168, 169, 174, 181, 186, 188, 196; dialectic 168–9; evolution of 168, 182–3, 185, 198; as form of positivism 66–9, 70, 188; human *167*; as Kuhnian paradigm 177–8, 180–2, 184, 185, 186, 191; as Lakatosian research paradigm 185; Piagetian 166, 174; pragmatic *167*; radical 63–4, 166, *167*, 169; social, see social constructivism; trivial *167*
contemporary philosophers of science 4, 34, 35
continual critical appraisal of data 22, 23
contradictions x, 1, 8, 9, 27, 29, 31, 33, 124, 169, 183, 184, 185; Lederman-Osborne 161–3, 165, 191
controversies ix, 1, 5, 6, 9, 21, 31, 32, 33, 164, 181, 182, 188, 190; generalization and 81, 84; role in physical science textbooks 23–4, 25–7, 29; *see also* historical controversies
Cook, T.D. 95
Cooper, L. 4
cooperation and collaboration 150, *160*, 161
Copernican astronomy 25, 36
Cortéz, R. 74, 89, 104, 106
Coulomb's law 38

counter-arguments 33, 44
counterexamples 15
creativity 3, 8, 9, 127, 147, 150, 151, 155–6, 159, *160*, 164, 190, 191
critical historical perspective 5
critical testing 150, 159–60, 161, 163–4, 164–5, 191
cross-domain analogies 36
Crowther, J.G. 23, 24
cultural embeddedness of science 28, 150
cultural milieu 6
curriculum 1, 3, 4, 17, 20, 149, 186, 188; chemistry 11, 12–13; constraints 3; content 5; materials 33; new 33; traditional 1, 33
Cushing, J.T. 151, 165
cutting-edge research 29, 139; *see also* speculative experiments

Dalton, J. 28; model of the atom 28, 40, 55; theory of multiple proportions 40
Darrigol, O. 193
Darwin, C. 36, 43
De Berg, K.C. 147–8
De Posada, J.M. 115, 130
developmental stage theory 37, 38, 43–4, 73, 108, 167
Dewey, J. 35, 48
dialectic constructivism 168–9
Dibar, M.C. 152, 170, 174
disciplinary matrix 14
displacement thesis 15, 36–7, 45, 47, 48, 54, 87, 113, 168
dispute in science 20, 32
diversity of scientific thinking 150, *160*, 161, 164, 191
Dobson, C.B. 103, 107, 112
Dobson, K. 150
Donner, J. 43
Duffy, T.M. 186
Duhem-Quine thesis 34, 72, 97, 156, 170, 198

education, problems in, *see* problems in education
educational research: alternative approaches to methodology in 9–10, 49–70; generalization in 71–85, 90, 96–8, 101; mixed (integrative) methods 46, 48, 50, 53, 62, 65–6, 69, 91, 133, 154, 173; multiple data sources in 91–2, 133, 154, 173; philosophy of 34; qualitative 49, 50, 71–85, 86–102; quantitative 35, 49, 50; warrants 34
Ehrenhaft, F. 132, 139, 140; *see also* Millikan-Ehrenhaft controversy
Einstein, A. 9, 48, 117; photoelectric equation 8; special relativity theory 41
Eisner, E.W. 86
electromagnetism, Maxwell-Lorentz theory 31, 36
electrons 7, 41, 42, 86, 87, 117, 128, 145, 197
elementary electrical charge 129; *see also* Millikan-Ehrenhaft controversy
Elliot, J. 50
empirical evidence, *see* experimental data
empiricism 12
empiricist epistemology 2, 3, 8, 128
empiricist perspectives 3–4
English as a foreign language 55
english teachers, elaboration of hypothesis and prediction *106*, *111*
epistemic subject 44, 73, 84, 101, 168, 196
epistemic transition 44
epistemological beliefs/frameworks 12, 47, 119, 122, 123, 125
epistemological realism 4
epistemology 2–3; empiricist 2, 3, 8, 128; genetic 43, 73, 84, 167; positivist 3, 128
Epistemology of Science Teaching course 11, 12, 152–65, 169–86
Erickson, F.E. 50, 54, 66, 68, 74, 76, 89
evidence-based research 34, 45
experimental data (empirical evidence) 3, 9, 40, 41, 48, 97, 149; interpretation of 21, 72, *160*, 191; *see also* alternative interpretations; reliability of 20, 21; role of in physical science textbooks 22–8; under-determination of theories by 8, 72, 97, 156, 164, 191; validity of, *see* validity
experimentalism, classical 47
experimenticism 41
experiments 19; historical embeddedness of 45; speculative, *see* philosophy of speculative experiments
Eysenck, H.J. 43

Falconer, I. 7, 128
fallibility 3
falling bodies, law of 13, 14
false science 5
falsificationism 44, 54, 150
falsified history 21
Fernández, I. 152, 170
Feynman, R. 38
Fleck, L. 18
fractional charges 7, 42–3, 97, 117, 197; *see also* quarks
free fall, law of 13, 37, 145
Freidson, E. 71, 96
Freire, P. 50, 64
Friman, P.C. 45
Fuller, S. 27
Furió, C. 115, 130
Furio-Mas, C. 28

Gage, N.L. 1

Galilean idealization 9, 13–14, 48, 196; in social sciences 43–4
Galileo 43; ideal law 13; law of free fall 13, 37, 145
Galison, P. 4, 196
Gallegos, J.A. 155, 170, 174
García, J.J. 115, 130
Garfield, E. 95
Gaskins, S. 50
Geiger, H. 7, 23, 26
Gell-Mann, M. *167*
general relativity 4
generalization: controversies and 81, 84; internal/external 71, 81, 83–4, 96; in qualitative research 10, 70, 71–85, 88, 90, 96–8, 101–2, 189, 196; in quantitative research 10, 71–2, 81, 98, 101–2
genetic epistemology 43, 73, 84, 167
Gholson, B. 36, 37, 46
Giere, R.N. 4, 9, 35, 39, 48, 187
Gil-Pérez, D. 152, 168, 170, 175, 186
Gilbert, J. 28
Giroux, H.A. 50
Glasersfeld, E.V. *167*, 169, 186
Glasson, G.E. 50, 57, 74, 89, 166, 169, 170, 180
Gooday, G. 5, 16
Gould, S.J. 150
Gower, B. 38
gravitational theory 15, 38–9
Greenbowe, T.J. 115, 130
Guba, E.G. 46–7, 48, 49, 71, 74, 76, 85, 86, 88, 89, 95, 133, 154, 173
guiding assumptions 1, 9, 23, 32, 33, 35, 128, 188

Handbook of Mixed Methods in Social and Behavioral Research 46
Hanson, N.R. 103
hard-core of belief 15, 23, 25, 114, 122, 123, 124, 125, 128, 190
health and sports participation 109–10
heat phenomena 114, 117, 197
Heilbron, J.L. 116
Heisenberg, W.C. *167*
Henry, J. 26
hermeneutic phenomenology 170
heuristic principles 7, 18, 128, 129
heuristic/explanatory power 26–7
historical controversies 7–8, 11, 22, 23–4, 25–7, 114, 129, 134, 147–8; *see also* alpha particles; cathode ray experiments; Millikan-Ehrenhaft controversy
historical embeddedness of experiments/ theoretical arguments 45
historical episodes x, 4, 23–8, 33, 39–43, 44, 138, 150, 158, 161, *167*, 188, 190, 197, 198
historical milieu 3, 127
historical reconstructions 1, 9, 20, 32, 33, 36, 44, 139–40, 188, 197, 198
historical studies 4
history of science 7–8, 12, 16, 18, 19–20, 21, 22–3, 25, 49, 86–7, 124, 138, 162, 165, 193; conceptual change in 116–17, 120; perpetual flux in 5
Hodson, D. 168
Holton, G. 8, 9, 25, 26, 35, 42, 48, 117, 128, 144, 195, 196
human constructivism *167*
hybrid atomic models 28
hybrid research design 50, 74, 91, 133, 154–5, 173
hypotheses 9, 19, 22, 35, 38, 159, 160, 164; auxiliary 15; formulation and understanding of 10, 99, 100, 101, 102, 103–12, 189–90; proliferation of in education 33; rival 31, 33, 45, 54, 188; testing 150, 194
hypothetico-deductive reasoning 10, 53, 104, 106, 111

idealization 101, 125, 139–40, 145, 146, 147, 164, 189, 197; Aristotelian 102, 189; Galilean 13–14, 37, 43–4, 48, 196; research methodology of 16
ideas-about-science 11, 149–65, 190–1
imagination 3, 9, 127, 150, 155
imperative of presuppositions 22, 23, 38, 39, 42, 43, 44, 45, 48, 193
in-service teachers 2, 3, 186
inclined plane experiment 13
incommensurability thesis 36, 37, 46–7, 48, 91, 98, 113, 133, 154, 173, 177, 191
inconsistent foundations 3, 25, 27, 36, 127, 174
individual differences 44, 168
inductivism 150, 156, 174
infallibility 3
Infeld, L. 117
inference 150
Inhelder, B. 108
inquiry activities 12
instructional strategies 33
integrated historical teaching 5
integration of qualitative and quantitative methods 48, 53, 62, 65–6, 69, 84, 188; *see also* mixed methods (integrative) research
interpretation of data, *see* experimental data (empirical evidence), interpretation of
Investigation in the Teaching of Chemistry courses 11, 114–25, 129–48

James, W. 35, 48
Jenkins, E. 29, 32
Jiménez, E. 152, 170, 174

Johnson, J. 50
Johnson, M.A. 108
Johnson, R.B. 34, 35, 37, 46, 48, 91, 133, 154, 173
Jones, R.C. 25
Journal of Mixed Methods Research 46
Justi, R. 28

Karplus, R. 57
Kaufmann, W. 7, 41, 42, 86, 128
Keeves, J.P. 81
Kelvin dictum 22
Kerlinger, F.N. 50, 73, 88
Kerwin, M.L.E. 45
Kesidou, S. 33
kinetic theory 27, 114, 117, 129, 197
Kitchener, R.F. 13, 44, 84, 167
Klassen, S. 4
knowledge: constitutedness/constitutiveness of 67; growth and meaning of 9, 50, 52–69, 74, 88, 115, 152, 169; *see also* scientific knowledge
Kousathana, M. 115, 130
Kuhn, D. 108
Kuhn, T. 116
Kuhn, T.S. 16, 17–18, 35, 45, 46, 49, 55, 72, 97, 174; and constructivism 168, 169; normal science 8, 9, 14, 18, 20, 21, 23, 29, 30–1, 32, 87, 148, 157–8, 164, 175, 177, 178, 188, 191, 194; paradigms 14–15 (displacement thesis 15, 36–7, 45, 47, 48, 54, 87, 113, 168; incommensurability thesis 36, 37, 47, 48, 98, 113, 177, 191)
Kuhnian approach to teaching 30, 32

Ladson-Billings, G. 43
Lakatos, I. 35, 46, 48, 49, 72, 94, 95, 97, 167, 169, 174; on *ceteris paribus* clauses 14; hard-core of beliefs 15, 23, 25, 114, 123, 128, 190; on Newton's law of gravitation 38; research programs 15–16, 19–20, 45, 185 (Bohr 25, 30–1, 137, 144; competing/rival 54–5, 113, 156; inconsistent foundations 25, 36); on tentative nature of science 26
Lakatosian approach to teaching 30, 32
Lakatosian methodology 15, 16
Lalik, R.V. 50, 57, 74, 89, 166, 169, 170, 180
LaMaster, S.U. 50, 57, 74, 89, 123, 166, 169, 170, 180
Larzelere, R. 45
Laudan, L. 9, 45, 174
Lavoisier, A. 39
laws 3, 103, 127, 145; and theories, distinctions and relationship between 150
Lawson, A.E. 103, 108
Le, A. 28
leaning tower of Pisa mythical experiment 13, 37
learning: constituted/constitutive nature of 57; psychology of 46, 176, 186
learning cycle 57, 58
Lederman, N.G. 3, 5, 6, 126, 127, 141, 145–6, 150, 151, 152, 156, 161–2, 163, 165, 170, 174, 191
Lee, E.R. 29, 42, 43, 132, 146
Lee, H.B. 73, 88
Leech, N.L. 46
Leite, L. 28, 115, 130
Lewin, K. 14
lightquanta 8
Lin, H. 152, 170, 174
Lincoln, Y.S. 37, 49, 51, 74, 89, 91, 95, 168; on authenticity in research 85, 88; on generalization 71, 76; on incommensurability of paradigms 46–7, 48, 91, 133, 154, 173; on mixed methods research 47, 91, 133, 154, 173; concept of paradigm revolution 35–6; on qualitative paradigm 86; on use of quantitative data in qualitative research 88
Linn, M.C. 140
Liston, D. 1–2
logical positivism 12–13
Louden, W. 166
Loving, C.C. 4, 168, 169
Lynch, J.M. 5, 16

McComas, W.F. 127, 132, 163
Machamer, P. 21, 114, 182
Mackenzie, J. 35
McMillan, B. 4
McMullin, E. 14, 140, 164
Malthus, T. 36
Marfo, K. 87
Marín, N. 115, 118, 124, 152, 170, 174, 186
Marquit, E. 128
Marsden, E. 7, 23
Martínez, A. 50, 74, 76, 89, 152, 170, 174, 186
mathematics teachers, elaboration of hypothesis and prediction *106*, *111*
Matthews, M.R. 67, 99, 102, 130, 146, 152, 168, 170, 174, 186
Maxwell, J.A. 50, 51, 71, 74, 76, 88, 89, 95, 96
Maxwell, J.C. 25, 134; kinetic theory of gases 27; simplifying assumptions 27
Maxwell-Lorentz theory of electromagnetism 31, 36
Mayer, M.E. 86
mechanics, laws of 27, 39
Medawar, P.B. 195
Mellado, V. 152, 170, 174
memorization of science content 2, 33, 188
Mendeleev, D. 9, 27–8, 40–1, 48
Merton, R.K. 22

metacognition 117, 119, 120, 125
metasubject 44, 73, 168
methodological guideline 4
methodologists 46
methodology: alternative approaches to, in educational research 9–10, 49–70; Lakatosian 15, 16; problem situation and 2, 48, 61–2, 101, 188, 189, 196, 197, 198
methodology courses 2
Methodology of Investigation in Education courses 130; alternative appoaches to methodology in educational research 9–10, 49–70; generalization in research 73–84; hypotheses and predictions 104–12; qualitative methodology 88–102
Metz, D. 4
Michell, J. 43
Michelson-Morley experiment 41
Millar, R. 2
Miller, S.M. 87
Millikan, R.A. 9, 48, 132, 137, 140, 144, *167*; laboratory notebooks 26, 196; *see also* Millikan-Ehrenhaft controversy
Millikan-Ehrenhaft controversy, and the oil drop experiment 7–8, 25–6, 42, 99, 100, 101, 102, 114, 117, 124, 134, 139, 144–5, 156, 170, 176, 196, 197
Mischel, T. 122
Mishler, E.G. 95
mixed methods 2, 33
mixed methods (integrative) research 9, 34–48, 188; as democratic alternative 47; in education 46, 48, 50, 53, 62, 65–6, 69, 91, 133, 154, 173
Monk, M. 19
Montero, M. 46, 50, 61, 74, 89, 92
Moore, M.T. 87
Moreno, L.E. 152, 170, 174, 186
Morley, E. 41
Moseley, H.G.J. 40
motivation of students 5
multiple data sources 91–2, 133, 154, 173
multiple proportions, theory of 40
Musgrave, A. 39, 49

National Association for Research in Science Teaching (NARST) 2
National Research Council (NRC) 34
National Science Education Standards 19, 126
National Science Foundation 2
nature of science (NOS) 5–7, 28, 149, 157–8, 174, 186, 194; as empiricist epistemology 2–3; myths with respect to understanding of 131–2, 134–6, 147, 190; students understanding of 3, 6–7, 20, 21–2; teachers' understanding of 6, 11, 129–48, 151, 190, 192–3
negative heuristic 15, 16, 137, 140, 144, 190, 193
Nelson, M.W. 87
new science curriculum 33
Newell, A. 45
Newton, I. 9, 43, 48; hypothesis explaining the gas laws 27; law of gravitation 15, 38–9; laws of mechanics 27, 39; three laws of dynamics 15; *Principia* 27
Newtonian physics 47
Newtonian theory 4
Niaz, M. 4, 21, 23, 24, 25, 26, 27, 31, 38, 44, 50, 51, 73, 74, 84, 89, 104, 106, 114, 115, 117, 118, 128, 130, 139, 150, 152, 168, 170, 174, 175, 186, 193, 194
Nobel Laureates 4
Nola, R. 166
normal science 8, 14, 18, 19, 20, 21, 22, 23, 29, 30–1, 32–3, 87, 148, 157–8, 164, 168, 175, 177, 178, 188, 191–5
normal science education (NSE) 29
Novak, J.D. 44, 166
Nuffield Foundation 20
Null Hypothesis Significance Testing Procedure (NHSTP) 35

objective nature of science 4
objectivist realism 4
objectivity 6, 37, 52–3, 72, 97, 98, 127, 141, *142–3*, 144–5, 147, 190, 196
observation 3, 4, 67–9, 127, 145, 150, 196–7; participant 53, 66, 75, 90, 92, 98–9, 101, 102, 189, 197; as theory-laden 6, 34, 35, 66, 69, 94, 101, 103, 127, 166, 174, 189, 196; understanding of 104, 106
oil drop experiments, *see* Millikan-Ehrenhaft controversy
O'Loughlin, M. 108
Olson, J.K. 163
Onwuegbuzie, A.J. 34, 35, 37, 46, 48, 91, 133, 154, 173
Osborne, J. 19, 21, 127, 150–1, 152, 156, 159, 161–3, 165, 170, 174, 186, 191
Otero, J.C. 115, 117, 117–18, 130

Padilla, K. 28
paradigm revolution 35–6
paradigm wars 2, 34, 91
paradigms 1, 3, 6, 14–15; coexistence of competing 54, 87, 168; displacement thesis 15, 36–7, 45, 47, 48, 54, 87, 113, 168; emergent 36, 45; incommensurability of 36, 37, 46–7, 48, 91, 98, 113, 133, 154, 173, 177, 191
participant observation 53, 66, 75, 90, 92, 98–9, 101, 102, 189, 197
Pascual-Leone, J. 9, 13, 37, 44, 48, 73, 167, 168, 169, 193, 198

Pasteur, L. 43
Paul, J.L. 87
Pauli, W. *167*
pedagogy 2
peer pressure 1
peer review 3, 23, 127, 146
Peirce, C.S. 35, 48
Pera, M. 21, 182
periodic law of chemical elements 27–8, 40–1
Perkins, D.N. *167*, 185
Perl, M.L. 9, 29, 42–3, 48, 97, 132, 139, 140, 144, 145, 146, 148, *167*
perpetual flux 5
perspectival realism 187
Peshkin, A. 86
Petrucci, D. 152, 170, 174
Phillips, D.C. 4, 12, 34, 43, 45, 187, 188, 195, 196, 198
Philosophical Magazine 23, 86
philosophy of speculative experiments 29, 42–3, 97, 132, 139–41, 146–7, 148, 197
Phlogiston theory 39
photoelectric effect experiments 8
photosynthesis 6
Physical Review 26
physical science textbooks, *see* chemistry textbooks; physics textbooks
physics teachers, elaboration of hypothesis and prediction 105, *106*, 110, *111*
physics textbooks 8, 23, 24, 25, 27, 28, 128
Piaget, J. 9, 48, 70, 98, 167, 186, 193, 198; developmental stage theory 37, 38, 43–4, 73, 108, 167; epistemic subject 44, 73, 84, 101, 168, 196; generalization of research 72–3, 83, 84, 189, 196; genetic epistemology 43, 73, 84, 167; psychological subject 44, 84, 196
Piagetian constructivism 166, 174
Pinch, T. 21, 33
Planck, M. 25, 134
Polanyi, M. 72, 97, 174
Popper, K.R. 4, 9, 10, 31, 167, 169, 174
positive heuristic 15, 16
positivism 9, 10, 12–13, 40, 53, 85, 157, 166, 174, 187; classic 12; logical 12–13; social constructivism as form of 66–9, 70, 188
positivist epistemology 3, 128
positivists 35, 44, 46–7
Posner, G.J. 113, 124, 192
post-structuralism 170
postpositivists/postpositivism 34, 46–7, 187, 198
practicing researchers, methodologists need to catch up with 46, 185–98
pragmatic constructivism *167*
pragmatism 35, 48
pre-service teachers 2
predictions 10, 102, 103–12, 150, 159, 160, 164, 189–90
presuppositions 1, 9, 15, 22, 23, 29, 31, 33, 35, 128, 147, 190; imperative of 22, 23, 38, 39, 42, 43, 44, 45, 48, 193
problem situation, and methodology 2, 48, 61–2, 101, 188, 189, 196, 197, 198
problems in education 92, 94, 101; quantitative research and 99–101
progress in science *see*, scientific progress
progressive problem shifts 15, 73, 168, 193
progressive transitions in understanding of nature of science 11, 129–48, 190, 192–3
Project 2061 31, 19, 33
protective belt of auxiliary hypotheses 15
psychoanalysis 45
psychological subject 44, 84, 196
psychology 114, 167, 168, 176; behavioral 45; cognitive 9, 37, 43–4, 45, 48, 72–3, 186, 193, 196, 198; of learning 46, 176, 186
Putnam, H. 176

qualitative data 10, 35
qualitative methods 10, 50, 61–2, 65–6, 88; *see also* integration of qualitative and quantitative methods
qualitative research 2, 10, 34, 35; authenticity of 10, 75, 85, 88, 90, 95–6, 98–9, 101, 189; in education 49, 50, 71–85, 86–102; generalization in 10, 70, 71–85, 88, 90, 96–8, 101–2, 196; paradigm 36, 87–8, 91, 133, 154, 173; use of quantitative data in 88
qualitative understanding 14
quantitative data 9, 10, 23, 35, 48; and its statistical treatment 94–5; use in qualitative research 88
quantitative imperative 22, 38, 39, 43, 44, 45, 48, 193
quantitative methods 2, 10, 50, 62, 65–6, 88; *see also* integration of qualitative and quantitative methods
quantitative research 10, 34; in education 35, 49, 50; generalization in 10, 71–2, 81, 98, 101–2, 189; paradigm 36, 87, 91, 133, 154, 173; and problem solving in education 99–101; samples 81; validity in 85, 88, 95, 189
quantitative understanding 14
quantum of action 25, 116, 124, 134, 170
quantum mechanics 4, 28, 47, 54, 55
quantum theory 36
quarks 29, 42–3, 97, 100, 139, 145, *167*
questioning, science and 150, *160*
Quine, W.V.O. 72, 97, 164

radical constructivism 63–4, 166, *167*, 169
random samples 10

randomized controlled experiments, as gold standard 34
rational arguments 3, 127
Ratnesar, N. 35
realism: epistemological 4; objectivist 4; perspectival 187
record keeping 3, 127
reform efforts 2, 9, 19, 29, 31, 32
relativity theory 4, 41
reliability of evidence 20, 21
replicability 3, 127, 163
research methodology, *see* methodology
research programs: Bohr 15, 16; competing/rival 15, 19–20, 35, 50, 54–5, 94, 113, 114; dynamics of 44; Guba and Lincoln 35–6, 46–7; hard-core of 15, 23, 35; inconsistent foundations of 25, 27, 36; Lakatosian, *see under* Lakatos; mixed methods (integrative) 9, 34–48; negative and positive heuristic 15, 16; rival 15, 19–20, 35, 54–5, 94, 113, 114
research traditions 45
Rey, J. 39
rhetoric of conclusions 33, 129, 137
Rigden, J.S. 14
rival hypotheses/theories 3, 31, 33, 45, 54, 126, 127, 166, 174, 188
rival research programs 15, 19–20, 35, 54–5, 94, 113, 114
rivalries 9, 21, 32
Rodríguez, M.A. 23, 24, 25, 27, 139, 150
Roseman, J.E. 33
rote learning 2, 12
Russo, S. 29
Rutherford, E. 9, 24–5, 26, 28, 40, 48, 114, 132, 140, 144; alpha particles experiments 7, 23–4, 55, 116, 124, 144, 170, 197; model of the atom 5, 23, 26, 28, 30–1, 55, 116, 134, 139, 144, 161–2, *167*

Sadler, T.D. 6
Sale, J.E.M. 46
Saloman, G. 51, 61, 92
sampling 81; random 10
Sanger, M.J. 115, 130
Scharmann, L.C. 126, 152, 170, 174
school science 11, 32, 150
Schrödinger, E. *167*
Schrödinger's equation 28
Schwab, J.J. 18, 19, 33, 128, 129, 137
science curriculum, *see* curriculum
science education 11–12, 18, 20, 113, 114, 149; constructivism in 11–12, *167*, 168, 177–8, 182, 184, 185
science in the making ix, x, 1, 8, 19, 21, 30, 31, 32, 114, 125, 187, 188, 191–5, 197, 198
scientific community: critical appraisal of data 22, 23; *see also* peer review; and determination of research validity 95, 96
scientific knowledge ix, 3–4; cooperation and collaboration in development of 150, *160*, 161; creative and imaginative nature of 150, 159, *160*, 164; empirical nature of 150; historical development of 150, 159, *160*, 164, 191; social and cultural embeddedness of 28, 150; tentative nature of 1, 3, 6, 9, 11–12, 26, 32, 138, 144, 149, 157, 165, 166–7, 169, 182, 186, 187, 191, 192, 194, 198; theory-laden nature of 150, 155
scientific method 3, 29, 32–3, 52–3, 69, 72, 100, 102, 103, 125, 132, 148, 151, 157, 161–3, 188, 190, 196; chemistry teachers' understanding of 141, *142*, 145, 146, 147; and critical testing 150, 159–60, 161, 163–4, 164–5, 191; inevitable use of 151; myth of 29, 150, 151, 165, 191, 194; and nature of science (NOS) 126–7; representation in textbooks 124, 132, 137–9, 146, 192, 193
scientific papers 195–6
scientific practice 32, 168, 188, 190, 194
scientific progress 174, 185, 193, 195; ahistoric account of 29; constructivist interpretations of 176, 177, 184, 191; controversial nature of 9, 22, 23, 50, 72, 88–9, 100, 114, 181, 182; dynamics of 2, 4, 5, 23, 164, 176, 186, 187, 192; Kuhnian account of 49, 87; rival/competitive theories and 3, 126, 127
secondary schools, chemistry curriculum 11, 12–13
Segre, M. 37
Shayer, M. 108
Shulman, L.S. 45, 49–50, 51, 74, 91, 133, 154, 168
Siegel, H. 18
silenced voices 47–8
Sills, D.L. 22
Silver, M. 29
simplifying assumptions 14, 15, 125, 139, 146, 197
skepticism 3, 127
Slater, M. 5
Smeyers, P. 35
Smith, M.U. 126, 152, 170, 174
smoking and cancer 109
social constructivism 53, 58–61, 63, 66–9, 76, 123, 166, 167, 170, 179–80, 184, 188, 191; critical appreciation of 64, 70, 191; as form of positivism 66–9, 70; as Kuhnian paradigm 180–2, 184, 185, 191, 195
social embeddedness of science 28, 150
social milieu 3, 127
social sciences 43–4, 46, 49–50

socio-political context in scientific research 28
Solano, I. 152, 170, 174
Solbes, J. 115, 130
Sommerfeld, A. *167*, 193
spanish teachers, elaboration of hypothesis and prediction *106*, *111*
Spearman, C. 43
speculations 1, 3, 9, 28, 31, 33, 127, 147, 190
speculative experiments, philosophy of 29, 42–3, 97, 132, 139–41, 148, 197
sports participation and health 109–10
Stahl, G. 39
Stanford Linear Accelerator Center 42
Stanley, J. 95
statistical controls 10
Stern, O. 116
Stigler, S.M. 22
Stinner, A. 19 138
Strike, K.A. 113, 192
students: cognitive abilities 6; cultural milieu 6; epistemological views 12, 119, 122, 125; motivation 5; understanding of nature of science 3, 6–7, 20, 21–2, 31; window into the mind of 57, 58, 76
Stuewer, R.H. 14
sub-electrons 7, 42, 117, 197
subjectivity 14–15, 36, 97

Taber, K.S. 185
Tashakkori, A. 46, 47
Taylor, S.J. 50, 66
teacher education research 1–2
teacher training 2, 166–7; in-service 2, 3, 186; innovating 2; pre-service 2, 3
teaching approaches: English as a foreign language 55; Kuhnian and Lakatosian 30, 32
teaching science as practiced by scientists x, 194, 196, 198
teaching strategies 55–8
Teddlie, C. 46, 47
Teichman, J. 138
temperature 117
tentative nature of science 1, 3, 6, 9, 11–12, 26, 32, 138, 144, 149, 157, 165, 166–7, 169, 182, 186, 187, 191, 192, 194, 198
textbook history 21
textbooks 3, 4, 8, 9, 17–20, 186, 188; as pedagogical vehicles 8, 18, 31; *see also* chemistry textbooks; physics textbooks
Thagard, P. 36
theme generation 57–8
theories 3, 9, 19, 33, 45, 127, 188; construction of, as complex and creative process 151, 155–6; heuristic/explanatory power 26–7; historical embeddedness of 45; inconsistent foundations 3, 36, 127, 174; and law, distinctions and relationship between 150; rival 3, 126, 127, 166, 174; tentative nature of 26, 103; under-determination of 8, 72, 97, 156, 164, 191
Thompson, J. 165
Thomson, J.J. 9, 28, 40, 48, 87, 114, 132, 140; alpha particles experiments 7, 23–4, 116, 124, 134, 170, 197; cathode ray experiments 7, 23, 30, 41–2, 86, 128–9, 134; model of the atom 5, 24, 26, 30, 55, 134, 139, 161–2, *167*
Thomson, W. 22
Tobias, S. 186, 187
Tobin, K. 50, 57, 74, 89, 123, 166, 169, 170, 180
Toulmin, S. 16, 174
traditional science curriculum 1, 33
Traver, M.J. 115, 130
triangulation 91, 92, 94, 133, 154, 173
trivial constructivism *167*
truth in science 4
Tsai, C.-C. 3–4, 8, 166–7
Tsaparlis, G. 115, 130

under-determination of scientific theories 8, 72, 97, 156, 164, 191

validity 10, 20, 21, 53, 95–6, 101; degrees of 85, 95, 96, 101, 189; interpretative 75, 90, 95, 101; *see also* authenticity; in quantitative research 85, 88, 95, 189
value-free research 35
Van Aalsvoort, J. 12–13
Van Berkel, B. 29
Vaquero, J. 108
variables, manipulation of 99, 100, 101, 102
Verifiability Principle 12
Vienna Circle 12
Views of Nature of Science Questionnaire, Form B (VNOS-B) 150
Vuyk, R. 73
Vygotsky, L. *167*

Waldegg, G. 152, 170, 174, 186
Wallace, J. 166
Waters, M. 28
Weinberg, S. 4
Welljams-Dorof, A. 95
Whitcomb, J.A. 1–2
Wiechert, E. 7, 41, 42, 86–7, 128
Wilson, D. 24
Wilson, K.G. 5, 16
window into the students' mind 57, 58, 76
Windschitl, M. 165, 194